Automotive Powertrain Science and Technology

A motor vehicle's powertrain consists of the components which generate power and enable it to move – its engine, exhaust system, transmission, drive shaft, suspension and wheels. Any automotive engineering student going beyond basic mechanics will need a sound knowledge of the mathematics and scientific principles, particularly calculus and algebra, which underpin powertrain technology. This textbook supports a series of courses, for instance BTEC unit 28 "Further Mathematics for Engineering Technicians", which is a requisite for a foundation degree in automotive engineering, and BTEC higher unit 25 "Engine and Vehicle Design and Performance", without giving full coverage of automotive technology. It is a more focused companion to the author's *Automotive Science and Mathematics* 978-0-7506-8522-1, also published by Routledge.

Allan Bonnick, CEng MIMechE, trained and worked as an automotive technician before eventually becoming a principal lecturer and head of MV studies at Eastbourne College of Technology and Art. After retirement he joined the Technical Committee of the Institute of Road Transport Engineers, during which time he received the Mackenzie Junner Award for his work on electronically controlled systems on commercial vehicles.

Automotive Powertrain Science and Technology

Allan Bonnick

Routledge
Taylor & Francis Group

LONDON AND NEW YORK

First published 2020
by Routledge
2 Park Square, Milton Park, Abingdon, Oxon OX14 4RN

and by Routledge
52 Vanderbilt Avenue, New York, NY 10017

Routledge is an imprint of the Taylor & Francis Group, an informa business

British Library Cataloguing-in-Publication Data
A catalogue record for this book is available from the British Library

Library of Congress Cataloging-in-Publication Data
Names: Bonnick, Allan W. M., author.
Title: Automotive powertrain science and technology / by Allan
 Bonnick.
Description: Abingdon, Oxon ; New York, NY : Routledge, 2020. |
 Includes bibliographical references and index.
Identifiers: LCCN 2019031083 (print) | LCCN 2019031084 (ebook) |
 ISBN 9780367331139 (pbk) | ISBN 9780367331115 (hbk) | ISBN
 9780429318023 (ebk)
Subjects: LCSH: Automobiles—Power trains—Design and
 construction. | Engineering mathematics.
Classification: LCC TL260 .B66 2020 (print) | LCC TL260 (ebook) |
 DDC 629.2/4—dc23
LC record available at https://lccn.loc.gov/2019031083
LC ebook record available at https://lccn.loc.gov/2019031084

ISBN: 978-0-367-33111-5 (hbk)
ISBN: 978-0-367-33113-9 (pbk)
ISBN: 978-0-429-31802-3 (ebk)

Typeset in Sabon
by Apex CoVantage, LLC

Contents

Chapter 1

Gas laws and basic thermodynamics

1.1 Internal combustion engines

An internal combustion engine is a machine that converts heat energy into mechanical energy, and it does so in accordance with the first law of thermodynamics, also known as **Joule's law**, which states that "heat energy and mechanical energy are mutually convertible".

The Otto cycle, 4 stroke, engine is an example of a machine that converts heat energy into mechanical power. Immediately before (on the first stroke of the piston), air and fuel are induced into the cylinder. The air is used

1   To provide the oxygen for burning the fuel
2   To provide the gas that expands to drive the piston along the cylinder.

On the second stroke of the piston, a flywheel provides the energy that is used to compress the air/fuel mixture. Compression is required to ensure that the engine produces adequate power. At the end of this stroke, the fuel is ignited and the air in the cylinder is heated. On the third stroke of the piston, the heated air expands driving the piston, connecting rod and crank to provide the mechanical energy that makes the power output of the engine. The fourth stroke is required to remove burnt gases from the engine so that the cycle can start again.

1.2 Compression ratio

Figure 1.1(a) shows the piston at top dead centre and (b) at bottom dead centre, the volume of the air in the cylinder when the piston is at TDC, is the clearance volume V1. It is in this space that the air is heated. Figure 1.1(b) shows the piston at bottom dead centre, BDC; the additional volume is the swept volume of the cylinder, the total volume now being swept volume plus clearance volume. The compression ratio is: clearance volume plus swept volume ÷ clearance volume, in symbols,

$$r = \frac{Sv + Cv}{Cv}.$$

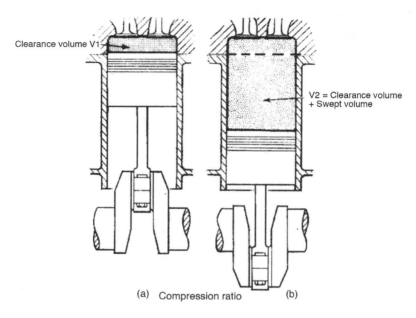

(a) Compression ratio   (b)

Figure 1.1 The compression ratio

In petrol engines, the compression ratio is approximately 8:1; in diesel engines, it is 14:1 and upwards.

Example 1.1

A single cylinder, 4 stroke engine has a bore diameter and stroke of equal size of 80 mm. Given that the clearance volume is 40 *cm³*, determine the compression ratio.

Solution example 1.1

$$r = \frac{Sv + Cv}{Cv}$$

Swept volume S_v = area of bore × length of stroke

The bore and stroke are of equal size at 8 cm.

$$S_v = \frac{\pi}{4} \times 8 \times 8 \times 8 = 402.2 \, \text{cm}^3$$

compression ratio $r = \dfrac{402.2 + 40}{40} = \dfrac{442.2}{40} = 11.05 : 1.$

1.3   Some terminology used in the work

STP stands for **standard temperature and pressure,** where the pressure is taken as 1.01325 bar and temperature of 0°C. It is used, in some applications, for reference purposes.

NTP, which stands for **normal temperature and pressure,** refers to atmospheric conditions on an average day. NTP is taken as a pressure of 1.01325 bar and 15°C. It is used in work such as engine testing where a standard for air flow, etc., is used.

Absolute pressure

In the normal working environment, pressures are recorded on gauges and meters and the pressures shown assume that the atmospheric pressure is zero. Absolute pressure is used in most work involving gases, and it is determined by adding atmospheric pressure to the gauge pressure.

Absolute temp

The gas laws are based on absolute temperatures. Absolute temperatures are measured in degrees Kelvin.

Absolute temperature in degrees K = temperature in degrees C + 273

System

In the study of heat engines, which is called thermodynamics, a system is a quantity of matter in which energy can be stored. There are two types of system:

a   **Closed system.** An example of this is the gas inside the engine cylinder on the various strokes.

The system shown in Figure 1.2 shows a simple engine cylinder where the piston is starting the compression stroke, the air in the system is contained by the cylinder and piston and these constitute the boundary. The process of compression is a typical **non-flow process.**

b   **Open system.** Most gas turbines operate as open systems, where there is a continuous flow of gas between the input and output.

In this simple representation of a gas turbine power plant (Figure 1.3), the air is first compressed and then passed into the combustion chamber, where it is heated. The heated gas then enters the high pressure stage of the turbine. The heated gas operates on the turbine blades to change heat into mechanical power at the output shaft.

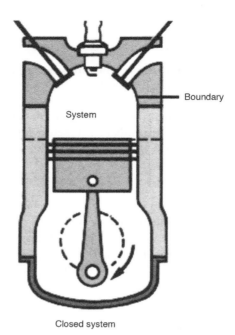

Boundary

System

Closed system

Figure 1.2 A closed system

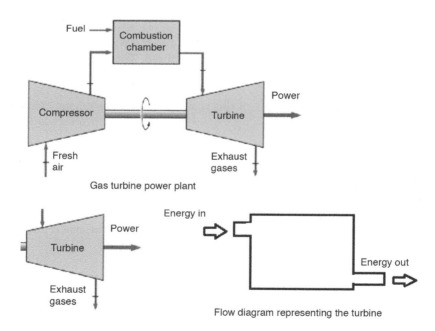

Fuel → Combustion chamber

Compressor

Power

Turbine

Fresh air

Exhaust gases

Gas turbine power plant

Turbine

Power

Exhaust gases

Energy in

Energy out

Flow diagram representing the turbine

Figure 1.3 Flows process in an open system

First law of thermodynamics

The first law of thermodynamics, also called Joule's law, states that "heat energy and mechanical work are mutually convertible".

The second law of thermodynamics

The second law of thermodynamics, also known the Clausius statement, states that "heat flows from a hot substance to a cold substance unaided, but it is impossible for heat to flow from a cold substance to a hotter substance without the aid of external work".

The second law means that a substance can absorb heat only from a source which is at a higher temperature than the substance itself. Where it becomes necessary to transfer heat from a cold body to one that is hotter than itself, mechanical energy must be employed, as in the case of air conditioning and refrigeration.

Atoms and molecules

An **atom** is the smallest particle of an element which can take part in a chemical reaction. A **molecule** is a finite group of atoms which is capable of independent existence and has properties that are characteristic of the substance of which it is the unit. A molecule may consist of one or more atoms, and this is shown in the chemical formula of a substance. For example:

H_2 (hydrogen) has two atoms to the molecule and is said to be diatomic.
CO_2 (carbon dioxide) has 3 atoms to the molecule, 1 of carbon and 2 of oxygen, and it is said to be triatomic.

Relative molecular mass (weight) is the mass of a molecule of a substance referred to that of an atom of Carbon 12.

Substance	Chemical symbol	Atomic mass	Number of atoms per molecule	Relative molecular mass
Hydrogen	H_2	1	2	2
Carbon	C	12	1	12
Nitrogen	N_2	14	2	28
Oxygen	O_2	16	2	32
Sulphur	S	32	1	32

1.4   The kgmol

The kilogram mole (kgmol) of a substance is the mass of the substance that has the same numerical value as the relative mass. For example, a kgmol of carbon has a mass of 12 kg.

Avogadro's hypothesis

Avogadro's hypothesis is "Equal volumes of gases at the same temperature and pressure contain equal numbers of molecules".

Using Avogadro's law (Figure 1.4), each container holds an equal number of molecules. If n is the number of molecules in each container, the *mass of gas in each = number of molecules × mass of each molecule*.

Applying the equation $pV = mRT$ to container 1

$$\frac{pV}{T} = nM_1R_1$$

Similarly for container 2, $\dfrac{pV}{T} = nM_2R_2.$

As $\dfrac{pV}{T}$ is a constant, $M_1R_1 = M_2R_2 = $ Constant

The constant MR is called the universal gas constant, and it has a value of 8.314 KJ/kmol.

If the molecular mass of a gas is known its characteristic constant, R may be calculated from the equation $MR = 8.314$; for example, oxygen which has M = 32.

$$\therefore R = \frac{8.314}{32} = 0.260\,\frac{kJ}{kgK}.$$

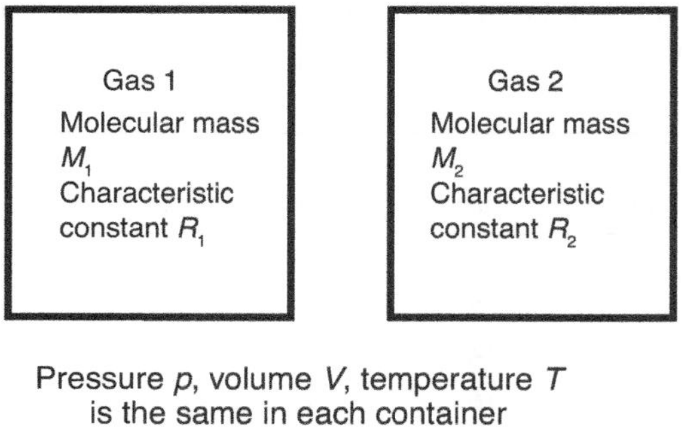

Gas 1	Gas 2
Molecular mass M_1	Molecular mass M_2
Characteristic constant R_1	Characteristic constant R_2

Pressure p, volume V, temperature T
is the same in each container

Figure 1.4 Avogadro

1.5　Dalton's law of partial pressure

Dalton's law states that the total pressure in a vessel that contains a mixture of gases is the sum of the partial pressures which each gas would exert if it occupied the space alone. This principle is applied in the design of zirconia type oxygen sensors, where the partial pressure of oxygen in the exhaust gas is compared with the partial pressure of oxygen in the atmosphere. The partial pressure of oxygen in atmospheric air is approximately 200 mbar and that of oxygen in the exhaust gas varies with the mixture strength, as shown in Figure 1.5.

1.6　Expansion and compression of gases

The heat energy that makes the engine operate is contained in chemical energy in the fuel. The combustion process that is part of the cycle of events on which the engine operates releases energy in the form of heat which then heats the air in the engine cylinder, causing it to expand doing work on the piston and engine mechanism, leading to power at the engine output shaft. The behaviour of gas (air) when heated, expanded and compressed is an important factor in the operation of an internal combustion engine, and the

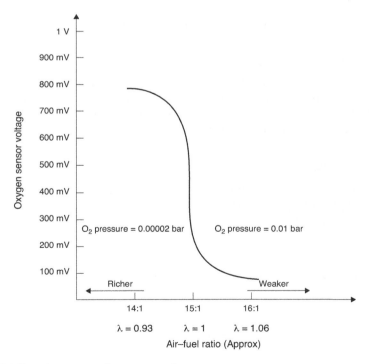

Figure 1.5 Partial pressure of oxygen in exhaust gas

various gas laws are considered in the next session. When dealing with gas laws, it is important to remember that temperatures are in degrees Kelvin ($°K = °C + 273$); pressures are absolute (gauge pressure plus atmospheric pressure); pressures are in Pascals (Pa); at $1N/m^2$, the Pa is a small measure; and the Bar at $1bar = 10^5$ Pascals, is normally used in calculations where heat engines are concerned.

Laws governing the behaviour of gases: Boyle's law (isothermal)

Boyle's law states that when a given mass m kg of gas is kept at *constant temperature*, its volume varies inversely as its absolute pressure. The law is expressed mathematically as $V \propto \dfrac{1}{p}$, or $pV = constant$. In effect this means that if the volume is halved, the pressure will be doubled. It is normally used in the form $p_1 V_1 = p_2 V_2$. This process is also known as an **isothermal** process because the temperature is constant.

Boyle's law (isothermal) compression or expansion is used in the form;

$$p_1 v_1 = p_2 v_2.$$

Pressures are in Pa absolute and volumes are in m^3.

Example 1.2

A volume of 5 m^3 of a perfect gas is contained in a cylinder at a pressure 140 kPa absolute. The gas is compressed until its pressure is 560 KPa absolute. Determine the volume of gas at the end of compression. The process takes place at constant temperature.

Solution example 1.2

First pressure $p_1 = 140 kPa$
Final pressure $p_2 = 560 KPa$
First volume $V_1 = 5m^3$
Final volume $V_2 = ?$

Solution 1.2 using Boyle's law

$$p_1 v_1 = p_2 v_2.$$

$$\frac{p_1 V_1}{p_2} = v_2 = \frac{140 \times 10^3 \times 5}{560 \times 10^3} = 1.25 m^3$$

$$V_2 = 1.25 m^3$$

Charles's law (constant pressure)

Charles's law states that when a given mass m kg of gas is kept at *constant pressure*, its volume varies directly as its absolute temperature.

This is expressed mathematically as $V \propto T$, $or \dfrac{V}{T} = \text{constant}$.

It is normally used in the form $\dfrac{V_1}{T_1} = \dfrac{V_2}{T_2}$.

Example 1.3

If 0.30 m³ of gas at a temperature of 50°C are heated at constant pressure until the volume is doubled, what will be the final temperature?

Solution

Using Charles's law:

$$\frac{V_1}{T_1} = \frac{V_2}{T_2}.$$

$$T_2 = \frac{v_2 T_1}{v_1} = \frac{0.6 \times 323}{0.3} = 646\,\text{K} = 373\text{C}$$

The characteristic gas equation

The characteristic gas equation is obtained by combining Boyle's law and Charles's law.

From Boyle's law, $pV = constant$, when T is constant.

From Charles' law, $\dfrac{V}{T} = constant$, when p is constant.

When combined, we have $\dfrac{p_1 v_1}{T_1} = \dfrac{p_2 v_2}{T_2}$.

For a constant mass of m kg, we may write,

$$\frac{pV}{T} = m \times constant$$

This constant is denoted by the letter R, and it is known as the characteristic gas constant. The characteristic gas equation is written in the form $pV = mRT$. Volume is in m³, pressure is in N/m², temperature is in degrees K, and mass m is in kg. In the case of air, $R = 0.287$ kJ/kg K.

Example 1.4

A cylinder contains 0.5 m³ of air at a temperature 165°C and pressure of 6.6 bar absolute. Calculate the mass of air in the cylinder.

Solution 1.4

Take R = 0.287 kJ/kg K.

$$pV = mRT$$

$$m = \frac{pV}{RT}$$

$$m = \frac{6.6 \times 10^5 \times 0.5}{0.287 \times 10^3 \times 438} = 2.63\text{kg}$$

1.7   Heating a gas

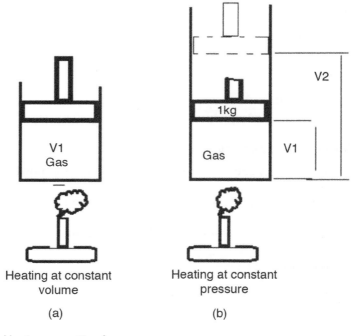

Heating at constant volume

(a)

Heating at constant pressure

(b)

Figure 1.6 Heating a quantity of gas

1.8   Conservation of energy

In heat problems the law of conservation of energy states that heat added *Q* is equal to work done by the gas *W* plus a change in internal energy *U*, in symbols;

$$Q = W + U$$

Heating gas at constant volume

Figure 1.6(a) shows a piston that is fixed in a cylinder so that the volume of gas remains constant. As there is no movement of the piston, no external work is done and all of the heat goes to increase the internal energy of the gas. The heat added to the gas increases its temperature and internal energy of the gas. The increase in internal energy is;

$$U_2 - U_1 = mC_v\left(T_2 - T_1\right)$$

where m is the mass in kg and C_v is the specific heat capacity of the gas at constant volume.

Heating gas at constant pressure

Figure 1.6(b) shows a quantity of gas at a pressure of p and volume V_1. As the gas expands, the piston is raised and the volume increases to V_2. At the same time the temperature of the gas rises from T_1 to T_2, the heat added is $Q = mC_p\left(T_2 - T_1\right)$ where C_p is the specific heat capacity of the gas at constant pressure.

The work done $W = p(V_2 - V_1)$, by the first law of thermodynamics the heat added is equal to the work done plus the change in internal energy.

$$Q = mC_p\left(T_2 - T_1\right) = p\left(V_2 - V_1\right) + mC_v(T_2 - T_1)$$

But $pV_2 = mRT_2$ and $pV_1 = mRT_1$

$$\therefore mC_p\left(T_2 - T_1\right) = mR\left(T_2 - T_1\right) + mC_v(T_2 - T_1)$$

Dividing through by $m\left(T_2 - T_1\right)$ leaves us with:

$$C_p - C_v = R$$

$$\frac{C_p}{C_v} = \gamma \text{ the adiabatic index for expansion or compression of the gas.}$$

Example 1.5

A vessel contains 0.15 kg of a gas at a temperature of 16°C and a pressure of 1.07 bar. The gas occupies a volume of 0.15 m³. Given that the specific heat capacity of the gas $C_v = 720$ J/kgK, calculate:

(a) the characteristic gas constant R
(b) the relative molecular mass of the gas M

(c) the specific heat at constant pressure C_p
(d) the adiabatic index.

Solution

(a) $pV = mRT$

$$\therefore R = \frac{pV}{mT} = \frac{1.07 \times 10^5 \times 0.15}{0.15 \times 289} = 370\,\text{J}/\text{kgK}$$

(b) Using the universal gas constant of 8314 J/kgmol, we have;

$$M = \frac{8314}{370} = 22.4$$

(c) $R = Cp - Cv \therefore C_p = C_v + R = 720 + 370 = 1.090\,\text{kJ/kgK}$

(d) $\dfrac{C_p}{C_v} = \gamma = \dfrac{1.090}{0.720} = 1.51$

Polytropic expansion or compression of a gas

The term polytropic applies to those processes where the index of expansion or compression is neither 1 nor γ.

A useful relationship between volume, temperature and pressure exists in

the form $\dfrac{T_2}{T_1} = \left(\dfrac{p_2}{p_1}\right)^{\frac{n-1}{n}} = \left(\dfrac{V_1}{V_2}\right)^{n-1}$, which is derived as follows:

Consider a mass of gas to change its state from p_1, V_1, 0_1 to p_2, V_2, T_2 according to the law $pV^n = C$.

$$pV^n = C \quad \therefore p_1 V_1^n = p_2 V_2^n \tag{1}$$

$$\frac{pV}{T} = C \quad \therefore \frac{p_1 V_1}{T_1} = \frac{p_2 V_2}{T_2} \tag{2}$$

From equation (2), $\dfrac{T_2}{T_1} = \dfrac{p_2 V_2}{p_1 V_1}$ $\tag{3}$

$$\left(\frac{V_2}{V_1}\right)^n = \frac{p_1}{p_2}$$

$$\therefore \frac{V_2}{V_1} = \left(\frac{p_1}{p_2}\right)^{\frac{1}{n}} = \left(\frac{p_2}{p_1}\right)^{-\frac{1}{n}}$$

Substituting in equation (3):

$$\frac{T_2}{T_1} = \frac{p_2}{p_1}\left(\frac{p_2}{p_1}\right)^{-\frac{1}{n}}, \text{ from which } \frac{T_2}{T_1} = \left(\frac{p_2}{p_1}\right)^{\frac{n-1}{n}}$$

By replacing pressures in equation (1) by using

$$\left(\frac{V_1}{V_2}\right)^n = \frac{p_2}{p_1} \text{ and substituting in equation (3) gives;}$$

$$\frac{T_2}{T_1} = \left(\frac{V_1}{V_2}\right)^n \times \frac{V_2}{V_1}$$

$$\frac{T_2}{T_1} = \left(\frac{V_1}{V_2}\right)^n \times \left(\frac{V_2}{V_1}\right)^{-1}$$

$$\frac{T_2}{T_1} = \left(\frac{V_1}{V_2}\right)^{n-1}$$

This then produces a very useful equation, which is;

$$\frac{T_2}{T_1} = \left(\frac{p_2}{p_1}\right)^{\frac{n-1}{n}} = \left(\frac{V_1}{V_2}\right)^{n-1}$$

Example 1.6

A quantity of gas at a temperature of 200°C is expanded to three times its original volume according to the law $pV^\gamma = C$, given that the specific heat capacities are $C_p = 1.03\,\text{kJ/kgK}$ and $C_v = 0.720\,\text{kJ/kgK}$, determine the final temperature of the gas.

Solution 1.6

$$\gamma = \frac{C_p}{C_v} = \frac{1.03}{0.720} = 1.43$$

$$T_1 = 473 \text{ K}$$

$$T_2 = ?$$

$$\frac{T_2}{T_1} = \left(\frac{V_1}{V_2}\right)^{n-1}$$

$$T_2 = T_1 \times \left(\frac{V_1}{V_2}\right)^{\gamma-1} = 473 \times \left(\frac{1}{3}\right)^{0.43} = 295 \text{ K}$$

Example 1.7

A certain compression ignition engine has a compression ratio of 16:1. At the beginning of the compression stroke, the gas pressure in the cylinder is 1.2 bar absolute.

The compression takes place according to the law $pV^{1.3}$. Calculate the gas pressure at the end of the compression stroke.

Solution 1.7

$$p_2 V_2^{1.3} = p_1 V_1^{1.3} \quad \therefore p_2 = p_1 \left(\frac{V_1}{V_2}\right)^{1.3}$$

$$\therefore p_2 = 1.2 \times \left(\frac{16}{1}\right)^{1.3}$$

$$p_2 = 1.2 \times 36.76 = 44.1 \, bar$$

Example 1.8

A certain compression ignition engine has a compression ratio of 22:1. At the beginning of the compression stroke, the pressure of the air in the cylinder is 1 bar and the temperature is 37°C. Compression is adiabatic where $\Upsilon = 1.41$. Calculate the temperature and pressure of the air in the cylinder, at the end of the compression stroke.

Solution 1.8

$$\frac{T_2}{T_1} = \left(\frac{V_1}{V_2}\right)^{n-1}$$

$$T_2 = T_1 \times \left(\frac{V_1}{V_2}\right)^{\Upsilon-1} = 310 \times 22^{0.41} = 310 \times 3.55 = 1100.5 \, K = 827.5 \, C$$

$$p_2 V_2^{1.41} = p_1 V_1^{1.41} \quad \therefore p_2 = p_1 \left(\frac{V_1}{V_2}\right)^{1.41} = 1 \times 22^{1.41} = 75.8 \, bar$$

1.9 Work done during expansion and compression

Work done during constant pressure expansion

In this case the pV diagram is a rectangle as shown in Figure 1.7.

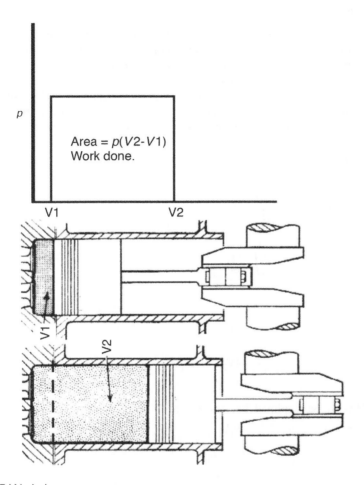

Figure 1.7 Work done at constant pressure

Example 1.9

A gas expands at a constant pressure of 15 bar as the volume increases from 0.005 m³ to 0.020 m³. How much work is done in this process?

Solution 1.9

Work done is equal to the area of the pV diagram.

$$Work\ done = p\left(v_2 - v_1\right) = 15 \times 10^5 \, N/m^2 \left(0.02 - 0.005\right) m^3 = 22.5 \, kJ$$

Polytropic expansion

Work done during polytropic expansion or compression – non-flow process in a closed system like an engine cylinder. A polytropic expansion or compression of gas is one in which the index of expansion is any value other than 1, or Υ which is the adiabatic index.

The area under the curve is the sum of the areas of the elemental strips of width dV, between the points p_1 and p_2 (as shown in Figure 1.8). In a polytropic process, pressure and volume are connected by the law $pV^n = c$.

$$\therefore p = \frac{c}{V^n}$$

$$\therefore \text{Work done} = \int_{V_1}^{V_2} p.dV.$$

$$= c \int_{V_1}^{V_2} \frac{dV}{V^n}$$

$$= c \left[\frac{V^{1-n}}{1-n} \right]_{V_1}^{V_2}$$

$$= \frac{c}{1-n} \left[V_2^{1-n} - V_1^{1-n} \right]$$

This can be simplified, because $c = p_1 V_1^n = p_2 V_2^n$, $p_2 V_2^n \times p_2 V_2^{n-1} = p_2 V_2$ and $p_1 V_1^n \times p_1 V_1^{n-1} = p_1 V_1$

$$\therefore \text{Work done } W = \frac{(p_2 V_2 - p_1 V_1)}{1-n}.$$

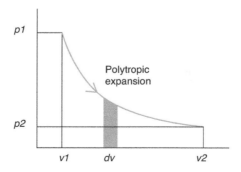

Figure 1.8 **Polytropic work**

This is normally given as $W = \dfrac{(p_1V_1 - p_2V_2)}{n-1}$

Example 1.10

A volume of 0.6 m³ of gas at a pressure of 7 bar and temperature of 170°C is expanded in a cylinder, with a moveable piston, to a pressure 1.5 bar. The expansion takes place according to the law $pV^{1.22} = C$. Find:

(a) the final volume and temperature of the gas
(b) the work done by the gas during the expansion
(c) the change of internal energy of the gas
(d) the heat flow across the cylinder walls.

Take $R = 0.287$ kJ/kgK, $C_v = 0.71$ kJ/kgK

Solution 1.10

(a) $p_1 = 7\,\text{bar}.\ p_2 = 1.5\,\text{bar}.\ V_1 = 0.6\,\text{m}^3$

$$V_2 = V_1\left(\frac{p_1}{p_2}\right)^{\frac{1}{n}} = 0.6\left(\frac{7}{1.5}\right)^{0.82} = 2.12m^3.$$

$$T_2 = \frac{p_2V_2}{p_1V_1} \times T_1 = \frac{1.5\times10^5 \times 2.12}{7\times10^5 \times 0.6}\times 443 = 335.4K$$

(b)   Work done $W = \dfrac{(p_1V_1 - p_2V_2)}{n-1} = \dfrac{7\times10^5 \times 0.6 - 1.5\times10^5\times2.12}{0.22}$

$$= 463.6 \text{ kJ}$$

(c)   Change of internal energy $= mC_v\,(T_2 - T_1)$. Use $pV = mRT$ to obtain m

$$m = \frac{p_1V_1}{RT_1} = \frac{7\times10^5 \times 0.6}{0.287\times10^3 \times 443} = 3.3kg. \therefore$$

$$(U_2 - U_1) = mC_v\,(T_2 - T_1) = 3.3\times0.71(335.4 - 443) = -252kJ$$

(d) $Q = W + (U_2 - U_1) = 463.6 - 252 = 211.6kJ.$ (heat added)

Adiabatic expansion and compression

An adiabatic expansion or compression is one in which there is no interchange of heat energy between the gas and its surroundings, it is in the form

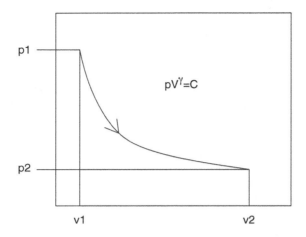

Figure 1.9 The work done in an adiabatic process, $W = \dfrac{(p_1 V_1 - p_2 V_2)}{\Upsilon - 1}$

$pV^{\Upsilon} = C$, where $\Upsilon = Cp/Cv$, which is the ratio of the specific heat capacities of the gas. Υ for air = 1.41 (Figure 1.9).

Example 1.11

An ideal diesel engine has a compression ratio of 14 to 1. At the start of compression the pressure of the air in the cylinder is 1 bar (abs) and the volume is 0.14 m³, at a temperature of 90°C. The air is compressed adiabatically. Calculate the work done on the air during the compression stroke. Take $\Upsilon = 1.41$.

Solution 1.11

First find p_2 using $p_2 = p_1 \left(\dfrac{V_1}{V_2}\right)^{1.41} = 1 \times 10^5 (14)^{1.41} = 41.3 \, \text{bar}$

$$
\begin{aligned}
\text{Adiabatic work done } W &= \frac{(p_1 V_1 - p_2 V_2)}{\Upsilon - 1} \\
&= \frac{1 \times 10^5 \times 0.14 - 41.3 \times 10^5 \times 0.01}{0.41} \\
&= -666 \, \text{kJ}
\end{aligned}
$$

The minus sign indicates work done on the air during compression.

Work done during isothermal expansion

An isothermal process is the same as Boyle's law where the expansion follows the law $pV = C$ (Figure 1.10). Following the reasoning used for the polytropic process, the work done is the sum of the areas of the elemental strips which is given by the integral:

$$W = \int_{V_1}^{V_2} p \, d_V$$

But $p = \dfrac{c}{V}$ and the integral becomes $W = \int_{V_1}^{V_2} \dfrac{c}{V} d_V$

$$= c \left[l_n V \right]_{V_1}^{V_2}$$

In this case $c = p_1 V_1$

$$\therefore \text{Isothermal work } W = p_1 V_1 l_n \frac{V_2}{V_1}$$

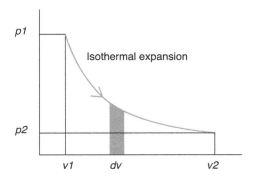

Figure 1.10 Isothermal work

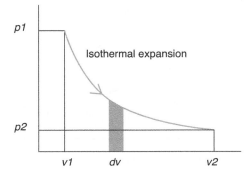

Figure 1.11 Isothermal work

Example 1.12

0.2 m³ of air at an absolute pressure of 12 bar and temperature of 353°K are expanded isothermally into a volume 0.8 m³. Calculate the work done during this process.

Solution 1.12

$$\text{Isothermal work } W = p_1 V_1 l_n \frac{V_2}{V_1} = 12 \times 10^5 l_n \left(\frac{0.8}{0.2} \right) = 12 \times 10^5 \times 1.39 = 1668 \, kJ.$$

The Carnot cycle

The Carnot cycle is based on theoretical principles that if an engine could be made to operate on them would result in an engine that would produce the *highest possible thermal efficiency of any heat engine*. Carnot was a French engineer/physicist who was working in the early part of the 19th century, before internal combustion engines had been produced.

Figure 1.12 shows that the Carnot cycle consists of four operations:

1   Isothermal expansion from 1 to 2
2   Adiabatic expansion from 2 to 3
3   Isothermal compression from 3 to 4
4   Adiabatic compression from 4 to 1.

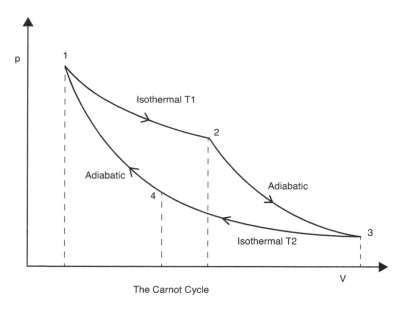

The Carnot Cycle

Figure 1.12 The Carnot cycle

The cycle commences at point 1 where the air in the cylinder is at a temperature of T_1. A source of heat is applied the end of the cylinder and the air expands isothermally to point 2, taking in heat as it does so. As the air expands, it does work on the piston. At point 2 on the diagram the source of heat is removed and the air expands adiabatically to point 3, doing further work on the piston and lowering the temperature to T_2. At point 3 a cold body at temperature T_2 is applied to the end of the cylinder and the piston reverses direction and compresses the air isothermally; during this stroke, heat is rejected to the cold body up to point 4 on the diagram. At point 4 the cold body is removed and compression takes place adiabatically to point 1, returning the temperature to T_1 in the process.

It should be noticed that heat was *taken in* during isothermal expansion from 1 to 2 only, and *heat was rejected* during isothermal compression from 3 to 4 only. There is *no* heat transfer during the two adiabatic processes.

Thermal efficiency of the Carnot cycle

Let p_1, V_1, T_1 be the condition of the air in the cylinder at point 1
Let p_2, V_2, T_2 be the condition of the cylinder air at point 3
Let $r = $ the ratio of isothermal expansion from 2 to 3
Let $r = $ the ratio of isothermal compression from 3 to 4

Both of these ratios must be equal for the cycle to function correctly.
Consider the cylinder to contain unit mass of air.

Heat supplied $= p_1 V_1 l_n r = R T_1 l_n r$

Heat rejected $= p_2 V_2 l_n r = R T_2 l_n r$

Work done = heat supplied − heat rejected $= R T_1 l_n r - R T_2 l_n r$

$$= R l_n r (T_1 - T_2)$$

$$\text{Efficiency} = \frac{work\ done}{heat\ supplied} = \frac{R l_n r (T_1 - T_2)}{R T_1 l_n r}$$

$$\therefore Efficiency = \frac{T_1 - T_2}{T_1}$$

This is the highest thermal efficiency that can be obtained for any engine cycle.

During the latter part of the 20th century the US government together with the Cummins engine company developed an engine known as the "low heat rejection LHR engine". The design made extensive use of ceramic materials, the aim being to obtain adiabatic compression and expansion, a high to temperature and a low final one, which would give it a high Carnot efficiency. Special lubricants were developed to cope with high operating temperatures, but the engine appears not to have received wide acceptance in industry.

Example 1.13

Calculate the Carnot efficiency of an engine that operates with a highest temperature of 2100°K and a lowest temperature of 500°K.

Solution

$$\text{Carnot efficiency} = \frac{T_1 - T_2}{T_1} = \frac{2100 - 500}{2100} = 0.76 = 76\%.$$

Summary of formulae used in this chapter.

$$pV = mRT$$

$$\text{Polytropic work done} = \frac{(p_1 V_1 - p_2 V_2)}{n - 1}$$

$$\text{Adiabatic work done} = \frac{(p_1 V_1 - p_2 V_2)}{\gamma - 1}$$

$$\frac{T_2}{T_1} = \left(\frac{p_2}{p_1}\right)^{\frac{n-1}{n}} = \left(\frac{V_1}{V_2}\right)^{n-1}$$

Change of internal energy $= mCv(T_2 - T_1)$

$$R = Cp - Cv \qquad \gamma = \frac{Cp}{Cv}$$

Heat added $Q = $ work done $+$ change of internal energy

$$\frac{p_1 V_1}{T_1} = \frac{p_2 V_2}{T_2}$$

$$\text{Compression ratio} = \frac{\text{Swept volume} + \text{clearance volume}}{\text{clearance volume}}$$

Self-assessment questions

1.1   A quantity of gas at a temperature of 20°C and pressure 1.3 bar (abs) occupies a volume of 2.5 m³. If the gas is now compressed until the volume is 1 m³ and pressure 7 bar (abs), calculate the final temperature.

Answer 1.1: 631°K, 385°C

1.2 During a short run the pressure and temperature of the air in a vehicle tyre are 2.5 bar and 48°C, respectively. After standing for a period of time the temperature of the air in the tyre drops to 16°C. Assuming that the volume remains constant, calculate the final gauge pressure. Take atmospheric pressure = 1.01 bar.

Answer 1.2: 3.16 bar (abs) = 2.15 bar (gauge)

1.3 Explain the term "isothermal process".
A quantity of gas has a volume of 0.0045 m³ at a temperature of 20°C and a pressure of 1.05 bar (abs). The gas is compressed isothermally to a final pressure of 6.30 bar (abs). Calculate the final volume.

Answer 1.3: 0.00075 m³

1.4 A cylinder contains 0.5 m³ of air at a temperature 165°C and pressure of 6.6 bar absolute. Calculate the mass of air in the cylinder. Take $R = 0.287$ kJ/kg K.

Answer 1.4: 2.63 kg

1.5 A vessel contains 0.15 kg of a gas at a temperature of 16°C and a pressure of 1.07 bar. The gas occupies a volume of 0.15 m³. Given that the specific heat capacity of the gas $C_v = 720$ J/kgK, calculate:

(a) the characteristic gas constant R
(b) the relative molecular mass of the gas M
(c) the specific heat at constant pressure C_p
(d) the adiabatic index.

Answer 1.5: (a) 370 J/kgK, (b) 22.4, (c) 1.090 kJ/kgK, (d) 1.51

1.6 A quantity of gas at a temperature of 200°C is expanded to three times its original volume according to the law $pV^\gamma = C$. Given that the specific heat capacities are $C_p = 1.03$ kJ / kgK and $C_v = 0.720$ kJ / kgK, determine the final temperature of the gas.

Answer 1.6: 295°K

1.7 A certain compression ignition engine has a compression ratio of 16:1. At the beginning of the compression stroke the gas pressure in the cylinder is 1.2 bar absolute. The compression takes place according to the law $pV^{1.3}$. Calculate the gas pressure at the end of the compression stroke.

Answer 1.7: 44.1 bar

1.8 A certain compression ignition engine has a compression ratio of 22:1. At the beginning of the compression stroke the pressure of the air in the cylinder is 1 bar (abs) and the temperature is 37°C. Compression is adiabatic where $\Upsilon = 1.41$. Calculate the temperature and pressure of the air in the cylinder, at the end of the compression stroke.

Answer 1.8: 827.5°C, 75.8 bar

1.9   A compression ignition engine draws in air at a pressure of 1 bar (abs) and compresses it adiabatically into the clearance volume before the fuel is injected. The stroke volume is 0.04 m³ and the clearance volume is 0.004 m³. Calculate:

(a) the pressure at the end of compression
(b) the final temperature, given an initial temperature of 30°C
(c) the compression ratio of the engine.

Answer 1.9: (a) 75.8 bar, (b) 791°K, (c) 11.1:1

1.10 One m³ of air at a pressure of 1.1 bar and temperature 15°C is com-pressed in a cylinder according to the law $pV^{1.25} = C$ in a cylinder until the pressure is 14 bar.

Given that R for air $= 0.287$ kJ/kgK and $C_p = 1.005$ kJ/kgK, calculate:

(a) the volume and temperature of the air at the end of compression
(b) the work done in compressing the air
(c) the change of internal energy
(d) the heat transfer across the cylinder wall.

Answer 1.10: (a) 0.131 m³, 207°C, (b) –293.6 kJ, (c) 183.3 kJ, (d) 110.3 kJ.

Chapter 2

Petrol engines

2.1 Ideal cycles of operation (spark ignition engines)

Detailed study of internal combustion engine behaviour normally takes place against a background of knowledge of some basic principles such as gas laws and the work done during expansion and compression of gases, such as air. Although these ideal cycles are based on theoretical processes which are impossible to achieve in practice, they form the principles against which the performance of practical engine cycles is compared.

2.2 The ideal Otto cycle

The ideal Constant Volume cycle (displayed in Figure 2.1), which is the basis of the Otto cycle, assumes a closed cylinder containing a set amount of air which is heated and cooled by an external source known as a heat sink.

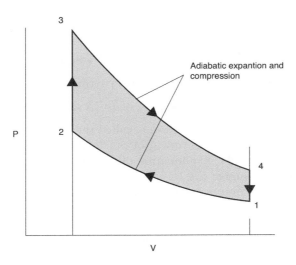

Figure 2.1 The Constant Volume cycle pressure volume diagram

Starting at point 1, the air is compressed adiabatically up to point 2, where the volume is small. At point 2 heat is added from an external source; this causes the pressure to rise to point 3. At point 3 the heat source is removed and the energy stored from the previous two processes causes the air to expand, doing work on the piston. At point 4 the heat sink is applied, and this causes the pressure to drop back to point 1.

2.3 Air Standard Efficiency (ASE)

The thermal efficiency of this cycle is called the air standard efficiency, and it is used as a guide to the thermal efficiency of practical engine performance.

Derivation of expression for ASE

The thermal efficiency of this ideal cycle is the amount of useful work divided by the amount of heat energy supplied. As no heat is added or subtracted during the adiabatic compression and expansion, the amount of useful work is the difference between the heat added during constant volume heating and extracted during cooling, namely:

$$mCv(T_3 - T_2) - mCv(T_4 - T_1).$$

The expression for thermal efficiency thus becomes;

$$\text{efficiency} = \frac{mCv\{(T_3 - T_2) - mCv(T_4 - T_1)\}}{mCv(T_3 - T_2)}.$$

$$= 1 - \frac{(T_4 - T_1)}{(T_3 - T_2)}.$$

Using the relation between temperature and volume; $\frac{T_1}{T_2} = \{\frac{v_2}{v_1}\}^{\gamma-1}$, the equation for efficiency can be produced in terms of compression ratio r because $r = \frac{v_1}{v_2}$. The thermal efficiency of this cycle is thus $1 - \frac{1}{r^{\gamma-1}}$. The efficiency derived from this equation is called the Air Standard Efficiency, and it is used to make an assessment of practical engines. Gamma (Υ) for air is approximately 1.41. Examination of the air standard efficiency equation shows that the higher the compression ratio r, the greater will be the efficiency and power output for a given amount of fuel. However, the graph in Figure 2.2 shows that above a compression ratio of 10:1, the increase in efficiency is quite small. In addition, there is the problem of detonation and preignition that sets a limit on the highest compression ratio that can be used.

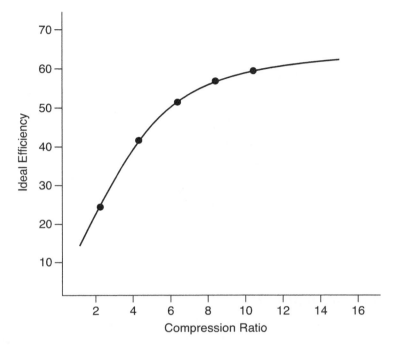

Figure 2.2 Limiting value ASE

Example 2.1

An engine operating on the constant volume Otto cycle has a compression ratio of 8.7:1. Calculate the air standard efficiency of this engine.

Solution

ASE at compression ratio of 8.7:1.

$$\text{ASE} = 1 - \frac{1}{8.7^{0.41}} = 1 - 0.41 = 0.59 = 59\%$$

This graph of ASE against the compression ratio shows that there is a large increase in ASE if the compression ratio is raised from 6:1 to 10:1, but close examination shows that after about 10:1, any further increase in the compression ratio leads to a smaller increase in ASE. This also applies to practical engines.

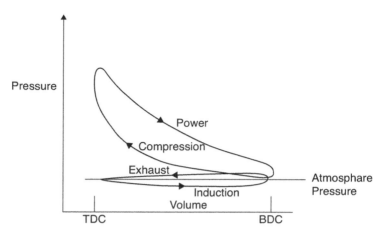

Figure 2.3 Indicator diagram 4 stroke

2.4   Indicated power

The indicated power of an engine is the power that is developed inside the engine cylinder. This is measured by means of an engine indicator, an instrument that records pressure inside the cylinder while the engine is working. The graph that is produced, called an indicator diagram, shows pressure against cylinder volume.

The area enclosed by the compression and power strokes represents the useful work done during one cycle, and it is necessary to determine this area and thus a figure for the useful work. An instrument called a planimeter is used to measure the effective area of the indicator diagram. The indicated mean effective pressure is then calculated by dividing the area of the diagram and then multiplying by the result by a constant that applies to the planimeter. The area of the diagram representing the induction and exhaust strokes is small and is normally disregarded in a test for maximum power.

Example 2.2

During a trial of a single cylinder, 4 stroke engine, an indicator diagram (Figure 2.4) is produced that has an area of 630 mm² and a base length of 70 mm. The spring constant for the planimeter is 90 kPa/mm. Calculate the indicated mean effective pressure. The area enclosed by the lines representing the induction and exhaust strokes is small and is normally disregarded in a test for maximum power.

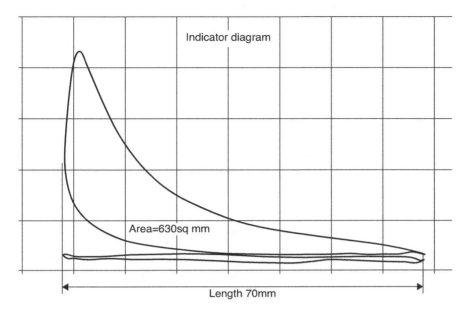

Indicator diagram

Area=630sq mm

Length 70mm

Figure 2.4 Indicator diagram for Example 2.2

Solution 2.2

Indicated mean effective pressure = area of diagram ÷ base length of diagram × indicator spring strength

Indicator spring strength = 90 kPa/mm of diagram height

Mean height of indicator diagram $= \dfrac{630}{90} = 7\,\text{mm}$

Indicated mean effective pressure (IMEP) $= 7 \times 90 = 630\,kPa = 6.3\,bar$

The indicated mean effective pressure is equivalent to an average pressure acting on the piston for the entire length of the power stroke.

Indicated power (IP)

Power = work done per second.

Indicated power = *plan*;

Where p = IMEP in Pa

l = length of stroke in metres

a = area of piston in m^2

n = number of working strokes per second

Example 2.3: indicated mean effective pressure

A 4 cylinder, 4 stroke engine develops an indicated mean effective pressure of 10 bar at a speed of 3000 rev/min. The bore and stroke are equal at 100 mm. Calculate the indicated power.

Indicated power $= plan$

IMEP p $= 10 \times 10^5$ N/m²

Area of piston a $= \dfrac{\pi d^2}{4} = \dfrac{3.142 \times 0.1 \times 0.1}{4} = 0.0076\, m^2$

Length of stroke l $= 0.1$ m

Working strokes n $= \dfrac{Number\, of\, cylinders}{2} \times \dfrac{rpm}{60} = 2 \times 50 = 100\, per\, second$

Indicated power $= 10^6 \times 0.1 \times 0.00786 \times 100 = 78600W = 78.6\, kW$

2.5   Indicated power by the Morse test

The difference between indicated power and brake power is largely due to friction and pumping losses. A fairly accurate value for indicated power can be obtained by the Morse test, in this test on a dynamometer a multi-cylinder engine is set to run at its optimum speed and the brake power is recorded. After the initial test when all cylinders are firing, further tests are conducted with one cylinder cut out. With a cylinder cut out, the dynamometer load is adjusted to restore the engine speed to what it was with all cylinders firing. The new dynamometer load is now used to calculate the brake power with one cylinder cut out. The power thus calculated is then subtracted from the power that was obtained with all cylinders firing, the difference between the two readings is the indicated power of the cylinder that is cut out. The procedure is repeated for all of the other cylinders so that the indicated power of the engine is obtained. The following example shows the method involved.

Example 2.4

The data in the table shows the figures obtained in a Morse test to determine the indicated power of a 4 cylinder engine. From these figures, determine the indicated power of the engine.

Cylinder cut out	None	Number 1	Number 2	Number 3	Number 4
Brake power (kW)	60	41	40	43	42
Indicated power (kW)		19	20	17	18

Indicated power of the four cylinder engine:

$IP = 19 + 20 + 17 + 18 = 74\, kW$

2.6   Brake power

The brake power of an engine is determined by testing on a dynamometer, which is also called a brake because it exerts a braking effect on the engine output shaft. In the simplest form of dynamometer shown in Figure 2.5, the rope wound round the circumference of the external flywheel exerts a frictional force on the rim of the flywheel, this force multiplied by the radius of the flywheel gives the torque that the engine is producing. The engine speed is also recorded as part of the test, and the data recorded permits calculation of the brake power. Modern dynamometers for high-speed multi-cylinder engines are normally electric or hydraulic.

Calculation of brake power:

$$\text{Torque} = (w - s) \times R$$
$$\text{Engine speed} = N\text{rev/min}$$

$$\text{Brake power} = \frac{2\pi TN}{60 \times 1000} \text{kW}$$

T is in Nm and N is the engine speed in rev/min.

Example 2.5

On a simple dynamometer test, an engine develops a torque of 150 Nm at a speed of 2000 rev/min. Calculate the brake power.

Solution

$$\text{Brake power} = \frac{2\pi TN}{60 \times 1000} = \frac{2 \times 3.142 \times 150 \times 2000}{60,000} = 31.42 \text{kW}.$$

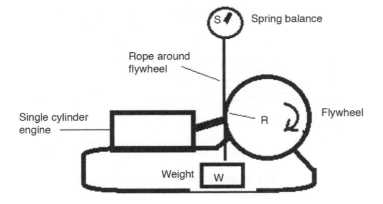

Figure 2.5 A basic dynamometer

2.7   Mechanical efficiency of an engine

Because of frictional and other losses, the brake power of an engine is considerably lower than the indicated power. The ratio of brake power to indicated power, $\frac{bp}{ip}$, is called the mechanical efficiency of the engine, and it is normally shown as a percentage. A value of around 80% is often quoted for mechanical efficiency of a multi-cylinder engine.

2.8   Brake mean effective pressure

The brake mean effective pressure (BMEP) of an engine is a figure that can be used when comparing the performance of one engine with another of similar design. It is normally quoted in an engine specification. In the engine map shown in Figure 2.6, the BMEP has a peak value of a little over 10 bar at an engine speed of 2200 rev/min. The other loops on this map show brake specific fuel consumption and the significance of the relation between BMEP and BSFC are considered in the section that deals with the transmission part of the drive train. The BMEP can be derived from the same formula used for calculating indicated power.

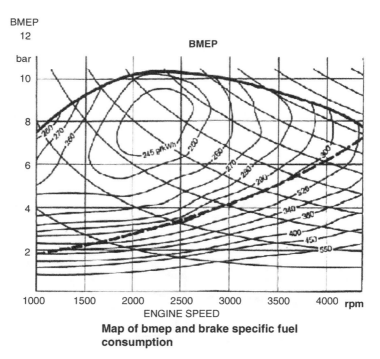

Map of bmep and brake specific fuel consumption

Figure 2.6 Brake mean effective pressure (BMEP)

Example 2.6

A 4 cylinder, 4 stroke engine develops a brake power of 50 kW at an engine speed of 2000 rev/min. The bore and stroke are equal at 110 mm. Calculate the BMEP.

Solution

$$bp = \frac{\text{BMEP} \times L \times a \times N}{60 \times 1000} \text{ kW}$$

$L = $ length of stroke $= 0.110$ m

Area of piston crown $= \frac{\pi d^2}{4} = 0.785 \times 0.11 \times 0.11 = 0.0095 \, m^2$

$N = bp = \dfrac{\text{number of cylinders}}{2} \times rpm = 2 \times 2000$ per min $= 4000$ per min

Transposing the equation bp gives BMEP $= \dfrac{bp \times 60 \times 1000}{L \times a \times N}$ Pa.

$$\text{BMEP} = \frac{50 \times 60 \times 1000}{0.110 \times 0.0095 \times 4000} = 718 \, kPa = 7.18 \, bar$$

2.9   Brake specific fuel consumption

Brake specific fuel consumption (BSFC) is a measure of the mass of fuel that is used by an engine in one hour while generating 1 kW of power. It is directly related to the brake thermal efficiency of the engine, and it shows how effective the engine is in converting fuel to power. The map in Figure 2.7 shows how BSFC varies with engine speed and torque, and it is a useful guide when comparing the performance of engines. The smaller the BSFC, the more effective the engine is in converting fuel to power.

Example 2.7

During a dynamometer test, a certain vehicle engine consumes 30 kg of fuel in one hour while it is developing 100 kW. Calculate the BSFC.

Solution

$$\text{BSFC} = \frac{\text{mass of fuel used in one hour}}{\text{brake power}} = \frac{30}{100} = 0.30 \text{ kg / kWh}$$
$$= 300 \text{ g / kWh}$$

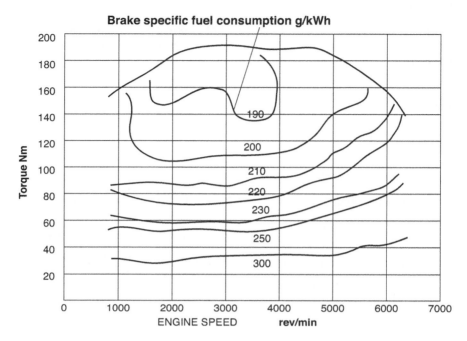

Figure 2.7 Graph of BSFC at various engine speeds

2.10   Brake thermal efficiency

The brake thermal efficiency of an engine is a measure of the effectiveness of the combustion and the way that energy in the fuel is converted into brake power. It is considerably less than the ideal air standard efficiency and a considerable amount of design work takes place in an endeavour to improve the practical brake thermal efficiency.

Brake thermal efficiency is calculated as follows:

Energy in brake power ÷ energy supplied in fuel in the same time

The brake power is normally quoted in kW, which is kJ per second, and fuel consumption is quoted in grams per hour.

Energy to brake power per hour = bp multiplied by 3600 kJ/h

Energy in fuel = calorific value multiplied by mass of fuel used per hour.

$$\text{Brake thermal efficiency} = \frac{\text{bp}\left(\dfrac{\text{kJ}}{\text{s}}\right) \times 3600}{\text{mass of fuel per hour} \times \text{cal value}}$$

Example 2.8

During a fuel consumption test on a dynamometer, an engine uses 15 kg per hour of fuel whilst developing a steady brake power of 60 kW. The calorific value of the fuel is 46 MJ/kg. Calculate the brake thermal efficiency.

Solution

$$BTE = \frac{60 \times 1000 \times 3600}{15 \times 46 \times 10^6} \times 100\% = 31.3\%$$

2.11   Volumetric efficiency

The volumetric efficiency of an engine is a measure of the effectiveness of the engine to draw in air for combustion.

Figure 2.8 shows how volumetric efficiency varies with engine speed. Diesel engines are not throttle governed, so the volumetric efficiency remains relatively high across the engine speed range; whereas in the petrol engine, the power output is controlled by the throttle and volumetric efficiency drops away at high engine speeds.

$$\text{Volumetric efficiency} = \frac{actual\ volume\ of\ air\ drawn\ into\ cylinder}{swept\ volume\ of\ the\ cylinder}$$

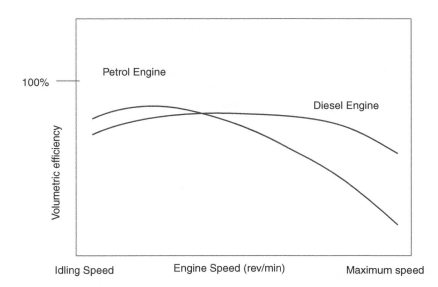

Figure 2.8 Volumetric efficiency V engine speed

Example 2.9

During a test on a dynamometer, a 4 cylinder, 4 stroke engine with a bore and stroke of 100 mm consumes air at a rate of 3 m³/min when running at 2200 rev/min. Calculate the volumetric efficiency of the engine.

Total swept volume of engine $\dfrac{\pi d^2}{4} l \times N$, where d is bore diameter, l is length of stroke and N is the number of cylinders.

$$\text{Total swept volume} = \frac{3.142 \times 0.10 \times 0.10 \times 0.10 \times 4}{4} = 0.00314\,m^3$$

Theoretical volume of air consumed per minute =

$$\text{total swept volume} \times \frac{\text{rev / min}}{2} = \frac{0.00314 \times 2200}{2} = 3.454 \text{ m}^3/\text{min}$$

$$\text{Volumetric efficiency} = \frac{\text{actual volume of air consumed / min}}{\text{theoretical value of air consumed / min}} \times 100\%$$

$$\text{Volumetric efficiency} = \frac{3.000}{3.454} \times 100 = 87\%.$$

For most engines the volumetric efficiency is of the order of 80%, and it is highly dependent on design of the induction system and valve timing. In turbocharged engines, the volumetric efficiency may exceed 100%.

The valve curtain area

Inlet valves are an important factor associated with volumetric efficiency. Two smaller inlet valves can provide a greater curtain area than one large valve, and this has a bearing on the amount of air (mixture) that can enter the cylinder in a given time. A feature of the valve known as the curtain area (Figure 2.9) determines the size of the aperture through which air (or mixture) enters the cylinder, and the variable in this area is the valve lift. There

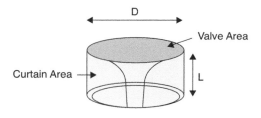

Figure 2.9 Valve curtain area

is a limit to the size of lift because there is little or no value in making the curtain area larger than the area of the valve port.

Valve curtain area $= \pi D L$

Ideally the curtain area should be equal to the area of the inlet port orifice.

$$\pi D L = \frac{\pi D^2}{4}$$
$$L = \frac{D}{4}$$

Relative efficiency

The relative efficiency of an engine is a method for showing how the indicated thermal efficiency of an engine compares with the theoretical air standard efficiency.

Example 2.10

An engine with a compression ratio of 8:1 working on the constant volume cycle returns an indicated thermal efficiency of 30%. Calculate the relative efficiency.

Solution 2.10

$$\text{Air standard efficiency} = 1 - \frac{1}{8^{0.41}} = 1 - 0.43 = 0.57 = 57\%$$
$$\text{Relative efficiency} = \frac{30}{57} = 52.6\%$$

2.12 Improving performance

Attempts to increase the power output of internal combustion engines is an ongoing process, and in recent years (such as 2017) much attention has been directed towards atmospheric pollution from NOx, particulate material (PM) and the greenhouse effect of CO_2. Two technologies, apart from exhaust after treatment, that are prominent in the 21st century are:

1   The Atkinson cycle
2   Direct petrol injection

2.13   The Atkinson cycle

In Figure 2.10, the area enclosed by the Atkinson pV diagram is larger than that for the Otto diagram. As this area represents work done on the piston per cycle, it follows that, for identical conditions, the theoretical Atkinson engine will produce a greater thermal efficiency. The Atkinson cycle consists of two adiabatic processes, a constant volume process, and a constant pressure process. A mass of air is contained in a closed cylinder and the heating and cooling of the air drives the piston. Heat is supplied to the enclosed air in a constant volume process, which is shown as 2 to 3 on the pV diagram. At point 3 the heat source is removed and the air in the cylinder expands adiabatically to point 4; this is the power stroke. At point 4 a cold source is applied outside the cylinder, and this causes the volume to decrease while the pressure remains constant. At point 2 the cold source is removed and the air is compressed adiabatically up to point 2. At point 2 the heat source is applied, raising the pressure to point 3, where the cycle repeats itself. It should be noted that the expansion (power) stroke is longer than the compression stroke, and this feature makes the thermal efficiency of an Atkinson slightly better that of a comparable Otto engine. Although Atkinson gas engines were used in the textile industry in the late 19th and early 20th centuries, the complicated mechanism that was necessary to give a longer power stroke was not mechanically reliable, and their use was discontinued. However, in the 21st century with increased concern about atmospheric pollution, renewed interest in the Atkinson engine has led to the development of engines that operate on the Atkinson-Miller principle, which emulates the Atkinson principle.

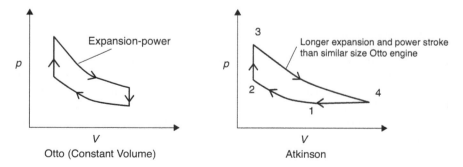

Figure 2.10 Atkinson V Otto

2.14 Thermal efficiency of Atkinson cycle

Because no heat is added during adiabatic expansion (3 to 4 on the PV diagram), and none is rejected during adiabatic compression (1 to 2 on the PV diagram), the thermal efficiency of an engine operating on the Atkinson principle is

$$\frac{\text{heat added} - \text{heat rejected}}{\text{heat added}} = \frac{mCv(T3-T2) - mCp(T4-T1)}{mCv(T3-T2)}$$

This can be reduced to the following equation in terms of the expansion and compression ratios where x is the expansion (power) stroke and r is the compression stroke:

Thermal efficiency $E = 1 - \gamma \dfrac{(x-r)}{x^{\gamma} - r^{\gamma}}$.

Example 2.11

The ideal efficiency E of an Atkinson engine with expansion ratio x of 9.6:1 and compression ratio r of is given as $E = 1 - \gamma \dfrac{(x-r)}{x^{\gamma} - r^{\gamma}}$. Calculate the efficiency of an Atkinson engine where r = 5:1, $\Upsilon = 1.41$ and $x = 9.6$:1.

Solution 2.11

Atkinson efficiency $E = 1 - \gamma \dfrac{(x-r)}{x^{\gamma} - r^{\gamma}}$

$$E = 1 - 1.41 \left[\frac{(9.6 - 5)}{\left(9.6^{1.41} - 5^{1.41}\right)} \right]$$

$$E = 1 - 1.41 \left\{ \frac{4.6}{24.3 - 9.7} \right\}$$

$$E = 55\%$$

Compare this value with ideal efficiency of an Otto engine operating at the same compression ratio of 5:1.

Otto engine ideal efficiency $= 1 - \dfrac{1}{5^{0.41}} = 0.48$, or 48%.

This shows that the ideal Atkinson engine will return a greater fuel efficiency than the ideal Otto engine, and it is this consideration that leads to designs such as Atkinson-Miller engines.

Example 2.12

Compare the ideal efficiency of a constant volume Otto engine with a compression ratio of 8:1 with an engine operating on the Atkinson cycle with a compression ratio of 8:1 and an expansion ratio of 16:1. Take Υ as 1.41.

Solution 2.12

$$\text{Otto, constant volume} = 1 - \frac{1}{8^{0.41}} = 1 - \frac{1}{2.35} = 1 - 0.43 = 0.57 = 57\%$$

Atkinson $r = 8$, $x = 16$

$$\text{Ideal efficiency} = 1 - \left[\frac{\gamma(x-r)}{x^\gamma - r^\gamma}\right] = 1 - \frac{1.41(16-8)}{16^{1.41} - 8^{1.41}} = 1 - \frac{11.28}{49.9 - 18.77}$$

$$= 1 - \frac{11.28}{31.1} = 0.64 = 64\%$$

2.15   The Atkinson-Miller engine

The Atkinson-Miller engine has the basic structure of a normal 4 stroke engine, the main difference being the way in which the power stroke is lengthened. In effect the power stroke and exhaust stroke are long and the compression stroke is shortened. The effective shortening of the compression process is achieved by leaving the inlet valve open for part of the compression stroke. During this process part of the charge from the induction stroke is returned to the intake manifold which is under pressure from the turbocharger.

Valve timing of Atkinson-Miller

A common use of the Atkinson-Miller principle is in the engines of hybrid vehicles where variable valve timing is also used. For power the traditional valve timing, where the inlet valve opens before TDC and closes a short period after BDC, is used; whereas for economy, the inlet valve is opened after TDC and closed 100 degrees or so on the compression stroke. This late closing of the inlet valve effectively shortens the compression stroke and also pushes a certain amount of the cylinder contents into the intake manifold. To counteract this, Atkinson-Miller engines are normally turbocharged. Because the combustion chamber space is small a relatively high compression ratio of about 12:1 is claimed for some engines.

2.16   Direct petrol injection

In the direct petrol injection system, fuel is injected directly into the cylinder in two ways: one is called homogenous injection mode, and the other stratified charge mode. A common rail fuel system operating at a pressure of approximately 120 bar is used to assist with thorough mixing of fuel and air and to allow some injection to take place on the compression stroke.

With homogenous charge, compression ignition fuel is *injected* during the intake stroke. A mixture of fuel and air (or other oxidizer) are compressed to the point of auto-ignition.

In stratified charge mode, injection takes place on the compression stroke, a few degrees before ignition occurs. The fuel is injected into the specially shaped combustion space in the piston crown. A feature of the combustion space design is that a circular volume of air surrounds the combustion region; this ensures a highly combustible mixture near the sparkplug, provides a degree of heat insulation for the combustion space metalwork and helps to prevent detonation. It is this *layering* of the combustible mixture that leads to the term *stratified charge*. The engine control system varies the operating mode to suit running conditions from a weak mixture, possibly 20:1 for light running conditions, to something like 12:1 for maximum power. The carefully designed intake manifold and combustion chamber permit high compression ratios of 10:1 or slightly more.

2.17 Exhaust gas recirculation

Exhaust gas recirculation (EGR) is a widely used strategy that aims to reduce NOx by lowering the combustion temperature to a figure somewhat lower than 1800°K, which is the temperature above which NOx is formed.

Exhaust gas recirculation is controlled by the engine management computer which brings EGR into operation when the engine has reached operating temperature. When conditions are satisfactory and the engine is not idling, EGR takes place after the inlet valve has opened and well before it is closed. Under load EGR is reduced until full load is reached at which point the EGR valve is closed to prevent further EGR.

2.18 Broadband O₂ sensor

Direct petrol injection engines operate with air-fuel ratios that vary from very weak (30:1 or 40:1, $\lambda > 1$), to rich (approximately 12:1, $\lambda < 1$) The broadband exhaust oxygen sensor is able to cope with this wide range of air-fuel ratios and is widely used in direct petrol injection and some diesel engines.

Self-assessment questions

2.1 A turbocharged 6 cylinder, 4 stroke engine with a bore and stroke of 100 mm returns a volumetric efficiency of 112% at 3000 rev/min. Calculate the volume of air per minute that the engine consumes.

Answer 2.1: 7.92 m³/min

2.2   A 4 cylinder, 4 stroke engine develops a brake power of 70 kW at an engine speed of 4200 rev/min. The bore and stroke are 90 mm and 100 mm respectively. Determine the brake torque and the BMEP at this speed.

Answer 2.2: 159.1 Nm, 5.24 bar

2.3   (a) The ideal efficiency E of an Atkinson engine with expansion ratio x of and compression ratio r of is given as $E = 1 - \gamma \dfrac{(x - r)}{(x^\gamma - r^\gamma)}$. Calculate the efficiency of an Atkinson engine where $r = 6{:}1, \gamma = 1.41$, $x = 10{:}1$

   (b) Describe how the Atkinson-Miller engine that is used in some modern engines differs from a true Atkinson engine.

Answer 2.3:

Atkinson efficiency $E = 1 - \gamma \dfrac{(x - r)}{(x^\gamma - r^\gamma)}$

$$E = 1 - 1.41 \left[\frac{(10 - 6)}{(10^{1.41} - 6^{1.41})} \right]$$

$$E = 1 - 1.41 \left\{ \frac{4}{25.7 - 12.51} \right\}$$

$$E = 0.573 = 57.3\%$$

2.4   During a steady speed dynamometer test a petrol engine develops an indicated power of 40 kW and consumes 9.3 kg of fuel in one hour. The calorific value of the fuel is 45 MJ/kg, and 9 kW are absorbed in friction and pumping losses. Calculate:

   (a)   the brake power
   (b)   the brake specific fuel consumption
   (c)   the mechanical efficiency
   (d)   the indicated thermal efficiency
   (e)   the brake thermal efficiency.

Answer 2.4: (a) 31 kW, (b) 0.3 kg/kWh, (c) 78%, (d) 34.4%, (e) 26.7%

2.5   A 6 cylinder, 4 stroke engine has a bore and stroke of 100 mm, and an indicator diagram shows that it develops an indicated mean effective pressure of 8 bar at an engine speed of 3600 rev/min. The compression ratio is 8:1. Calculate:

   (a)   the clearance volume
   (b)   the indicated power
   (c)   the air standard efficiency.

The relative efficiency given that the indicated thermal efficiency is 30%.

Answer 2.5: (a) 0.000113 m³, (b) 75.84 kW, (c) 0.574, (d) 0.523

2.6   With the aid of pV diagrams, explain how an engine operating on the Otto cycle produces an indicator diagram that differs from the indicator diagram for the ideal constant volume cycle.

2.7   The table shows the results obtained during a dynamometer test at a constant speed of 2000 rev/min on a 6 cylinder petrol engine.

Cylinder Cut out	None	No.1	No.2	No.3	No.4	No.5	No.6
Brake power kW	45	36.4	37.2	36.8	37.6	37	37
ip		8.6	8.8	8.2	7.4	8.0	8.0

Calculate:

(a) the indicated power
(b) the mechanical efficiency
(c) the total friction and pumping losses.

Answer 2.7: (a) 49 kW, (b) 92%, (c) 4 kW

2.8   A 4 cylinder, 4 stroke petrol engine with a bore and stroke of 100 mm and 120 mm respectively consumes atmospheric air at a rate of 2.75 m³/min at 2400 rev/min. If the BMEP is 8 bar and the brake specific fuel consumption is 0.3 kg/kWh, determine:

(a) the volumetric efficiency of the engine
(b) the brake power
(c) the mass of fuel used per minute
(d) the air-to-fuel ratio on a mass basis.

Answer 2.8: (a) 61%, (b) 60.3 kW, (c) 0.301 kg/min, (d) 11.83:1

2.9   On test an engine produces an indicated power of 40 kW and consumes fuel of calorific value 44 MJ/kg at a rate of 10 kg per hour. During the test the friction and pumping losses amount to 11 kW. Calculate:

(a) the brake power
(b) the brake specific fuel consumption
(c) the mechanical efficiency
(d) the indicated thermal efficiency
(e) the brake thermal efficiency.

Answer 2.9: (a) 29 kW, (b) 0.334 kg/kWh, (c) 72.5%, (d) 32%, (e) 23.7%

2.10 A 4 cylinder, 4 stroke petrol engine is running at a speed of 3000 rev/min. If the length of the stroke is 1.2 times the bore diameter, determine the diameter so that the engine will develop an indicated power of 40 kW when the indicated mean effective pressure is 8 bar.

Answer 2.10: Bore diameter = 81 mm

2.11 Cylinder filling
The amount of air or mixture that enters the cylinder on the induction stroke is affected by size of the valve curtain and the mean piston speed. The velocity of air in the induction stroke should be below sonic speed because at supersonic speed a condition known as choking occurs where uncontrolled pressure waves disturb the cylinder filling process.

Mean piston speed = 2 × stroke × rev per second

Example
Calculate the mean piston speed of a reciprocating engine with a stroke length of 110 mm when it is running at 6000 rev/min.

Mean piston speed = 2 × 0.110 × 100 = 22 m/s

2.12 In a 4 cylinder, 4 stroke engine with a bore of 100 mm and a stroke of 110 mm, the air velocity through the inlet valve port is 320 m/s. Calculate:

(a) the volume of air entering the cylinder each second when the engine is running at 6000 rpm
(b) the curtain area of the valve necessary to meet the required intake air velocity.

Answer 2.11: (a) 0.173 m³/sec, (b) 540 mm²

Chapter 3

Diesel (compression ignition) engines

The diesel cycle as originally conceived and used in marine and large industrial engines is also known as the constant pressure cycle because the heat addition takes place at constant pressure. These engines operated at low speeds of rotation. As technology became available in the 20th century, it became evident that the compression ignition principle could be applied to automobiles where the need for higher engine speed is required. As the technology developed it became evident that the constant pressure cycle was no longer suitable for road vehicles, and the dual combustion cycle was devised and is now used as the yardstick against which the operation of practical automobile engines is measured.

3.1 Diesel or constant pressure cycle

In the theoretical diesel cycle the pressure volume diagram is of the form shown in Figure 3.1. Working round the diagram, starting at point 1, air is compressed adiabatically – no heat is added to the air in the cylinder and none is rejected – up to point 2. At point 2 heat energy is added and the air in the cylinder expands, doing work on the piston. The volume increases to point 3 while the pressure remains constant; work is done on the piston during this phase. No further heat is added after point 3 and the air in the cylinder expands adiabatically, doing work on the piston up to point 4. At point 4 the remaining heat energy in the air in the cylinder is removed while the volume remains constant and the air in the cylinder is restored to its original state at point 1.

Because the compression and expansion processes are adiabatic during which no heat is added or rejected the air standard thermal efficiency of the diesel cycle is:

$$E = \frac{\text{heat added} - \text{heat rejected}}{\text{heat added}}$$

$$E = \frac{mC_p\left(T_3 - T_2\right) - mC_v\left(T_4 - T_1\right)}{mC_p\left(T_3 - T_2\right)}$$

$$= 1 - \frac{\left(T_4 - T_1\right)}{\gamma\left(T_3 - T_2\right)}$$

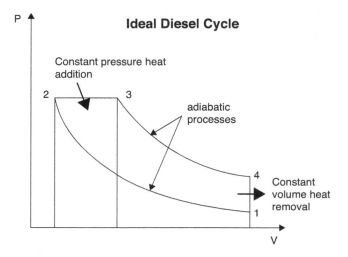

Figure 3.1 pV diagram for the diesel cycle

Example 3.1

A certain diesel engine has a compression ratio of 16:1, the air temperature in the cylinder at the start of compression is 45°C and at the end of combustion it is 1400°C. Calculate the air standard efficiency. Take $\gamma = 1.4$.

Solution 3.1

Temperature at the end of compression $= T_2$

$$\frac{T_2}{T_1} = \left(\frac{v_1}{v_2}\right)^{1.4-1}$$

$$\therefore T_2 = T_1\left(\frac{v_1}{v_2}\right)^{1.4-1}$$

$$= 318 \times 16^{0.4} = 964\,\text{K}$$

Find the cut-off ratio $\dfrac{V_3}{V_2} = \dfrac{T_3}{T_3} = \dfrac{1673}{964} = 1.74.$

Assuming that the cut-off ratio and the compression ratio represent units of volume:

$$\frac{T_4}{T_3} = \left(\frac{V_3}{V_4}\right)^{\gamma-1}$$

$$T_4 = 1673 \times \left(\frac{1.74}{15}\right)^{0.4} = 1673 \times 0.42 = 703\,\mathrm{K}$$

$$\mathrm{ASE} = 1 - \frac{(T_4 - T_1)}{\gamma(T_3 - T_2)}$$

$$T_1 = 318^\circ\mathrm{K}, T_2 = 964^\circ\mathrm{K}, T_3 = 1673^\circ\mathrm{K}, T_4 = 703^\circ\mathrm{K}$$

$$\therefore \mathrm{ASE} = 1 - \left(\frac{703 - 318}{1.4(1673 - 964)}\right) = 1 - \frac{385}{993} = 0.61$$

3.2   The dual combustion cycle

Large slow speed (100 rpm) diesel engines such as those used for marine propulsion or stationary power generation operate reasonably closely to the ideal diesel cycle. However, smaller, lighter weight engines that are necessary for road vehicles are required to operate at higher speeds and it is considered that the modified diesel cycle known as the **dual combustion cycle** more accurately represents the thermodynamic processes that occur in high-speed compression ignition engines.

In this cycle, the pV diagram for which is shown in Figure 3.2, the combustion takes place in two stages, one at constant pressure and one at constant volume. Working round the diagram from point 1, which corresponds to bottom dead centre on the compression stroke, the air in the cylinder is compressed adiabatically up to point 2. At point 2, heat is added, increasing the pressure, while the volume remains constant until point 3 is reached. At point 3 further heat is added at constant pressure, and expansion takes place up to point 4. From point 4 to point 5, the air in the cylinder expands

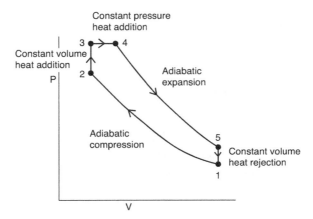

Figure 3.2 pV diagram for the dual combustion cycle

adiabatically. At point 5 the remaining heat is removed at constant volume to restore the air to its original condition ready to repeat the cycle. During the adiabatic compression and expansion processes no heat is added or rejected, and the work done during the cycle = heat added − heat rejected.

The air standard efficiency of the engine is operating on this cycle is;

Air standard efficiency = *heat added − heat rejected*
Heat added

$$E = \frac{mC_v(T_3 - T_2) + mC_p(T_4 - T_3) - mC_v(T_5 - T_1)}{mC_v(T_3 - T_2) + mC_p(T_4 - T_3)}$$

$$E = 1 - \left[\frac{(T_5 - T_1)}{(T_3 - T_2) + \gamma(T_4 - T_3)}\right]$$

Example 3.2

An engine working on the dual combustion cycle has a compression ratio of 16:1. In accordance with the ideal cycle, the compression and expansion are adiabatic with $\Upsilon = 1.4$. At the start of compression the temperature of the air in the cylinder is 318°K. Determine the following:

(a) The temperature at the end of compression.
(b) The air standard efficiency of the engine given the following temperatures: $T_3 = 1600°K, T_4 = 1850°K, T_5 = 800°K$.

Solution 3.2

(a) $\dfrac{T_2}{T_1} = \left(\dfrac{v_1}{v_2}\right)^{1.4-1}$

$\therefore T_2 = T_1\left(\dfrac{v_1}{v_2}\right)^{1.4-1}$

$= 318 \times 16^{0.4} = 964°K$

(b) $E = 1 - \left[\dfrac{(T_5 - T_1)}{(T_3 - T_2) + \gamma(T_4 - T_3)}\right]$

$= 1 - \left[\dfrac{800 - 318}{(1600 - 964) + 1.4(1850 - 1600)}\right]$

$= 1 - 0.49 = 0.51$, or 51%.

NOTE: When the constant pressure process (3 to 4 on the pV diagram) is very short, the dual combustion cycle is similar to the constant volume (Otto) cycle. For purposes of making comparisons between different types of engine, it is permissible to use the Otto cycle formula for ideal efficiency, ASE $=$ $1 - \dfrac{1}{r^{\gamma-1}}$. When this is done, it will be seen that the ideal thermal efficiency of any type of automotive compression ignition engine (CIE) is greater than that of an equivalent petrol engine. This is a factor that feeds through in practice where CI engine vehicles produce superior fuel consumption figures, and it is a consideration where fuel consumption is a major concern in the transport of goods and passengers. This advantage is somewhat overshadowed in recent years because of concerns about NOx and other exhaust emissions.

3.3   Comparison between theoretical and practical engine cycles

These ideal cycles reveal the thermal efficiency that can be achieved if ideal conditions prevailed. In the real world, ideal conditions do not exist; nevertheless designers of practical engines use the ideal cycles as a yardstick against which to compare the performance of working engines. Some of the factors that are taken into account when comparing practical engine cycles with theoretical ones are:

- The air and combustion gases in the cylinder are not perfect gases.
- Adiabatic processes are not realised in practice because after combustion, the working substance in the cylinder is no longer air but a mixture of gases, and it is virtually impossible to prevent interchange of heat between the cylinder content and the surroundings.

In spite of these difficulties, attempts have been made to develop adiabatic engines by utilising ceramic materials. An advantage claimed for them is that because of high operating temperatures, they are able to use a greater range of fuels.

3.4   The practical diesel (dual combustion) engine

Many of the differences between petrol engines and compression ignition ones arise from the combustion process. In order to cope with high pressures and the resultant stresses the component parts of CI engines are of more robust construction than comparable petrol engines which means that for a given power output, the CI engine is heavier. The table in Figure 3.3 shows some typical figures for weight-to-power ratio of a large truck engine.

The data in Figure 3.3 shows that the weight-to-power ratio of a large diesel engine is approximately 3 kg/kW.

Torque

As shown in Figure 3.4 a feature of the torque curve is that the torque is high at low engine speed and it remains reasonably flat throughout the speed range of the engine. This is a feature which, through engine mapping,

D2676 LF – comparison of key figures

	D2066 LF	D2676 LF	D2876 LF
Bore	120 mm	126 mm	128 mm
Stroke	155 mm	166 mm	166 mm
Capacity, 6 cylinders	10.52 l	12.42 l	12.82 l
Rating	320 – 440 hp	480 / 540 hp	480 / 530 hP
Torque	1600 – 2100 Nm	2300 / 2500 Nm	2300 / 2400 Nm
Cylinder spacing	154 mm	154 mm	158 mm
Weight of engine	975 kg / 440 hp E4	990 kg / 480 hp E4	1046 kg / 480 hp E3
Weight-to-power ratio	3.01 kg/kW	2.80 kg/kW	2.96 kg/kW

Figure 3.3 Weight-to-power ratio of engine

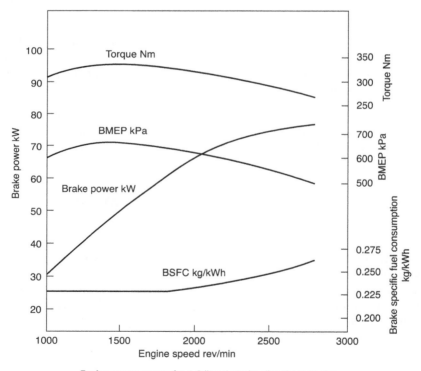

Performance curves for 1.6 litre 4 stroke diesel car engine

Figure 3.4 Performance details of a small diesel engine

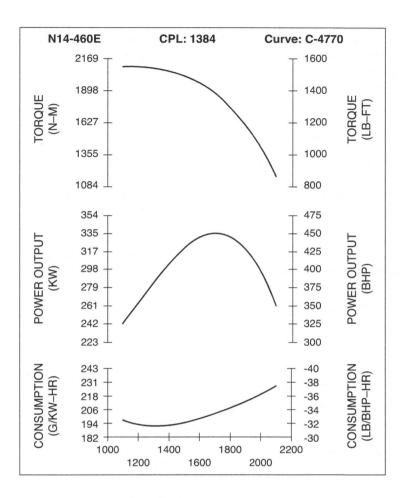

Cummins engine performance curves

Figure 3.5 Performance details of a Cummins diesel engine

can be engineered for a specific purpose as shown in Figure 3.5 which relates to a heavy haulage vehicle.

Brake mean effective pressure (BMEP)

The brake mean effective pressure of an engine provides a useful guide when comparing engine performance. It is independent of engine size and can be calculated from the torque curve, as shown in the chapter on spark ignition engines.

Brake power (bp)

This graph is obtained from data that is obtained during a dynamometer test. Diesel engine power is controlled by an engine speed governor the purpose of which is to prevent hunting and the top speed is limited to prevent the engine speed running out of control which could happen because the air supply to the engine is unlimited.

Brake thermal efficiency

The brake specific fuel consumption graph shows how the fuel consumption per unit of brake power varies with engine speed. It is a measure of the effectiveness of the engine in converting fuel into power. A low figure for BSFC means that the engine is efficient at converting fuel to power. Diesel engines generally have lower BSFC than petrol engines which shows up in the fuel consumption figures and lower CO_2 emissions. It should be noted that the minimum BSFC occurs when BMEP and torque are at their maximum value and this has bearing on the ideal operating range of the engine.

3.5 The fuelling map

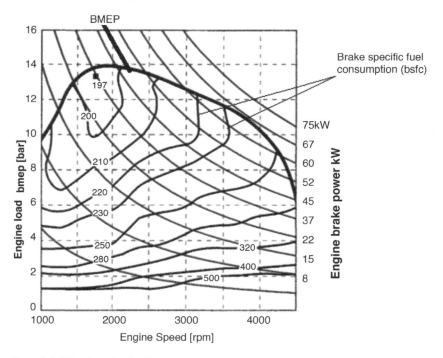

Figure 3.6 Diesel engine fuelling map

Example 3.3

The diesel engine fuelling map displayed in Figure 3.6 shows how brake specific fuel consumption varies with engine speed and load. The detail contained allows closer examination of the relationship between BMEP and BSFC; when the BSFC is 200 g/khr, the BMEP is at its highest value. From the data on the map it is possible to calculate the brake thermal efficiency as shown in the following example. In order to complete the calculation, it is necessary to know the energy content (calorific value of the fuel), in this case 45 MJ/kg.

Solution 3.3

$$\text{Brake thermal efficiency BTE} = \frac{\text{energy in brake power / sec}}{\text{energy supplied in fuel / sec}}$$

$$\textit{Energy supplied in fuel per second} = \frac{0.2 \times 45 \times 10^6}{3600}\text{J} = 2500\,\text{J}$$

$$\textit{Energy in brake power per second} = 1000\,\text{J}$$

$$\therefore \text{BTE} = \frac{1000}{2500} = 0.4, \text{ or } 40\%$$

Example 3.4

During a dynamometer test at steady speed, a compression ignition engine consumes 8 kg of fuel per hour and has a brake thermal efficiency of 30%. The calorific value of the fuel is 44.5 MJ per kg and the mechanical efficiency of the engine is 80%. Calculate the *indicated* power of the engine.

Solution 3.4

$$\text{Brake thermal efficiency} = \frac{\text{brake power} \times 3600}{\text{fuel mass} \times \text{cal val}}$$

$$\text{Brake power} = \frac{\text{BTE} \times \text{fuel mass} \times \text{cal val}}{3600} = \frac{0.30 \times 8 \times 44.5 \times 10^6}{3600}$$

$$= 29.7\ \text{kW}$$

$$\text{mech effy} = \frac{\text{bp}}{\text{ip}}$$

$$\therefore \text{ip} = \frac{\text{bp}}{\text{mech effy}} = \frac{29.7}{0.8} = 37.1\ \text{kW}.$$

3.6 Turbocharging

The power output of any internal combustion engine is ultimately limited by the mass of air that is introduced on the induction stroke. By forcing the air in on the induction stroke and increasing the mass of fuel to suit, the power output of the engine can be increased. This effect can be seen in Figure 3.7, where the characteristic performance curves for a naturally aspirated engine and a turbocharged one of the same capacity are compared. Curve (a) in Figure 3.7 represents the naturally aspirated engine and curve (b) the turbocharged one. The phase shown by D to E on the BSFC curve indicates the area where the turbo engine produces a better figure which arises because power and torque are higher and pumping losses are smaller. The brake

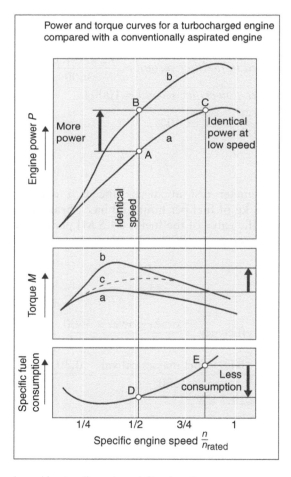

Figure 3.7 Turbocharge V naturally aspirated diesel engine.

power and torque figures for the turbocharged engine are considerably higher than those for the naturally aspirated one. At very low engine speeds the power and torque of the turbo engine are basically similar to those for the NA engine because there is very little energy in the exhaust stream, and this is the cause of turbolag. Although the increased power of the turbo engine is gained by using energy in the exhaust stream that would otherwise have been lost, some loss of power is caused by back pressure. Because the induction air is heated up during compression, it is less dense than cool air, which means that the mass of air is lower than it would be at atmospheric temperature. In order to overcome this disadvantage, it is common practice to cool the induction air in a heat exchanger.

Example 3.5

An atmospherically aspirated 4 cylinder, 4 stroke diesel engine with a bore diameter of 110 mm and a stroke of 120 mm develops a BMEP of 8 bar at an engine speed of 1500 rev/min.

(a) Calculate the brake power of the engine.
(b) The engine is modified to be turbocharged, and this raises the BMEP to 9.5 bar. Compare the power of the modified engine with that of the normally aspirated one.

Solution 3.5

(a) *Brake power* $= \dfrac{plaN}{60 \times 1000}$, where N = working strokes per min

$$\therefore \mathrm{bp} = \frac{8 \times 10^5 \times 0.12 \times 0.785 \times 0.11 \times 0.11 \times 3000}{60 \times 1000} = 45.6\,\mathrm{kW}$$

(b) $\mathrm{bp} = \dfrac{9.5 \times 10^5 \times 0.12 \times 0.785 \times 0.11 \times 0.11 \times 3000}{60 \times 1000} = 54.1\,\mathrm{kW}$

3.7 Combustion in the CI engine

Figure 3.8 shows two types of indicator diagram that were taken from a compression ignition engine with a compression ratio of 14:1 that was running at a steady speed of 2000 rev/min. Figure 3.8(a) shows the pressure volume diagram and (b) shows the pressure v crank angle diagram for the same dynamometer test. Point A is where fuel injection starts; it is followed by a delay period up to point B, where ignition begins. After point B there is a rapid increase in pressure in a few degrees of crank rotation followed by a short period where combustion continues without further rise in pressure.

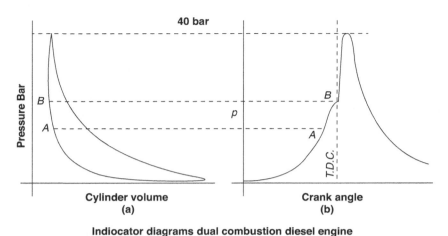

Indiocator diagrams dual combustion diesel engine

Figure 3.8 Phases of combustion in a diesel engine

These are the features which show a marked resemblance to the ideal dual combustion cycle.

Three phases of combustion

Sir Harry Ricardo carried out research in the early stages of high speed diesel engine development that showed that combustion takes place in three distinct phases. Figure 3.9 shows how cylinder pressure changes during these phases.

Phase 1. Point A on the diagram in Figure 3.9 is the point at which fuel injection starts. At this point the temperature of the compressed air in the cylinder is approximately 300°C which is sufficient to start combustion. The period from point A up to point B is known as the delay period, and it is the phase in which the droplets of fuel are being supplied with sufficient oxygen to support combustion. Engines are designed to keep the delay period as short as possible because it has an effect on the second phase of rapid pressure increase which is a cause of diesel knock. Turbulence is effective in reducing the delay period, as is turbocharging and the quality of the fuel which dependent on the Cetane rating.

Phase 2. From B to C on the diagram, there is a rapid rise in pressure as fuel injection continues. This phase approximates to the constant volume part of the ideal dual combustion cycle, and it occupies about 10 degrees of crank rotation. During this phase the combustion of the fuel already in the cylinder is completed.

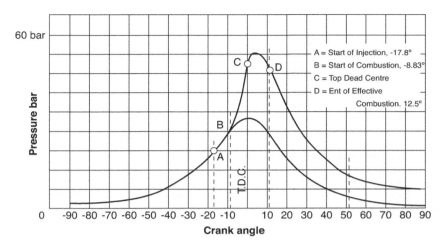

Figure 3.9 Phases of combustion

Phase 3. In phase 3 injection continues and the fuel burns as it leaves the injector. This phase approximates to the constant pressure part of the ideal cycle.

3.8   Steps taken to improve combustion

In order to achieve efficient combustion of the fuel, it is necessary to ensure that all of the fuel in the cylinder is supplied with an adequate amount of oxygen from the air in the cylinder, and this is achieved through turbulence.

Method of injection

Basically two methods of injection are used in automotive CI engines:

1   Direct injection, in which the fuel is injected directly into the combustion chamber.
2   Indirect injection, in which the fuel is injected into a pre-combustion chamber where it is mixed with air, so that combustion takes place as the burning gas enters the main combustion space. This is said to reduce the delay period with a resulting drop in diesel knock.

Compression ratio

Direct injection engines have a compression ratio 14:1 to 16:1 approximately, while indirect injections normally have higher compression ratios in the range of 20:1 to 30:1.

Steps taken to improve combustion

In order to achieve efficient combustion of the fuel, it is necessary to ensure that all of the fuel in the cylinder is supplied with an adequate amount of oxygen from the air in the cylinder. This is achieved through turbulence, of which there are two types:

1   Swirl induced on the induction stroke.
2   Squish type swirl created towards the end of the compression stroke.

3.9   Common rail and piezo injector

Modern diesel fuel injection systems make use of the common-rail system which operates at high pressures. The high fuel pressure coupled with accurate fuel injection several times per cycle that is possible with piezoelectric injectors leads to an improvement in fuel consumption and emissions.

Air-fuel ratio in compression ignition (diesel) engines

The power output of a diesel engine is controlled by the amount of fuel that is injected, from a very small amount of fuel for idling and low power, to a large amount for full power. The air supply for a diesel engine is not throttled, so the mass of air entering the cylinder does not vary greatly. This means that the **air-to-fuel ratio** of the diesel engine varies from very weak to quite rich and this has an effect on exhaust gas emissions, particularly NOx. The following excerpt from a report by Transport For London describes the effect of vehicle load and driving conditions on exhaust emissions, particularly NOx, PM and CO_2.

Drive cycle average emissions from heavy-duty Euro 6 vehicles un-laden and fully Laden.

Market segment	Fuel	Gross vehicle weight	NOx Test average 0% payload (g/km)	Test average 100% payload (g/km)	PM Test average 0% payload (g/km)	Test average 100% payload (g/km)	Test average 0% pay Load (g/ km)	CO_2 Test average 100% payload (g/km)
RN1 class iii LGV	Diesel	3500	0.494	0.682	0.002	0.001	256.9	290.1
N2 rigid HGV	Diesel	7500	0.71	0.357	0.003	0.003	315.1	470.85
N3 rigid HGV	Diesel	18000	2.714	0.511	0.006	0.007	672.45	921
N3 artic HGV	Diesel	40000	1.407	1.188	0.007	0.007	872.05	1797.45

Whilst levels of PM emission remain consistent regardless of payload, controlled by the diesel particulate filter (DPF), it is interesting to note that the NOx emissions are considerably lower in the fully Loaded condition for each vehicle type. This may be attributed to the increased engine exhaust temperatures on the laden vehicle allowing for more effective dosing of the SCR catalyst. In a number of cases, these cycle average emission Levels are almost as Low as those of diesel passenger cars, indicating the effectiveness of Euro VI at controlling NOx from heavy-duty engines, under the right conditions.

(Transport for London report, 2012)

Composition of diesel engine exhaust gas

Example 3.6

A traffic pollution survey for Transport for London found that lightly laden heavy diesel vehicles emitted greater amounts of NOx than heavily laden ones.

(a) Explain with reference to air-to-fuel ratios why you think that this could happen.
(b) Why is a broadband oxygen sensor used on diesel engine emission control systems?

The exhaust gases are as follows (see also Figure 3.10):

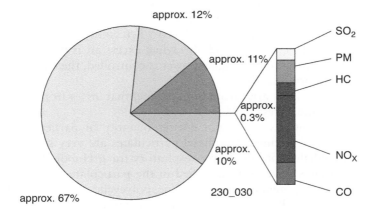

approx. 12%

approx. 11%

SO₂

PM

HC

approx.
0.3%

approx.
10%

NOₓ

approx. 67%

230_030

CO

Composition of exhaust emissions of desel engines

Figure 3.10 Diesel engine exhaust emissions (VW)

Solution 3.6

(a) At light load mixture is very weak which is likely to lead to higher NOx emissions.

(b) A broadband O_2 sensor is able to detect mixture strength when the mixture is weak.

> N_2: the air that is used for combustion contains about 77% nitrogen (N_2) by mass, so nitrogen forms a large percentage of the exhaust gas. It plays a part in the working of the engine because it expands when heated to provide the force that is exerted on the piston.
>
> CO_2: diesel fuel contains approximately 85% carbon and in the combustion process it is converted to carbon dioxide, CO_2.
>
> H_2O: diesel fuel contains approximately 15% hydrogen which is converted to water and steam.
>
> O_2: some oxygen that is not used in combustion will appear in the exhaust gas.

Pollutants: the small area to the right of the pie chart shows that diesel exhaust gas contains a small percentage of products of combustion that are considered to be harmful to human health. Note that although the percentage of pollutants is small, in areas where there is a concentration of vehicular traffic such as towns and cities, the build-up of pollutants in the atmosphere is a danger to health.

> CO: carbon monoxide appears in the exhaust gas when combustion is incomplete such as occurs when air and fuel have not mixed well and when excess fuel has been supplied.
>
> NOx: oxides of nitrogen are produced when combustion temperatures and pressures are high and also when excess air is present. Because of the way that diesel engine power is controlled, the diesel engine is prone to produce NOx.
>
> HC: hydrocarbons arise from unburnt fuel that arises from poor combustion and excess fuel.
>
> PM: diesel engines produce a greater quantity of particulate matter than petrol engines. The diesel particulates are very small and are easily inhaled into the lower respiratory tract. Hundreds of organic compounds have been identified in the particulate phase, many of which are potentially carcinogenic polycyclic aromatic hydrocarbons (PAHs). PM includes soot from incomplete combustion, minute metallic particles from engine component wear.
>
> SO_2: diesel fuel contains a small amount of sulphur which is converted to sulphur dioxide during combustion. A certain small amount comes from lubricating oil – more so in a worn engine.

Dealing with harmful emissions (pollutants).

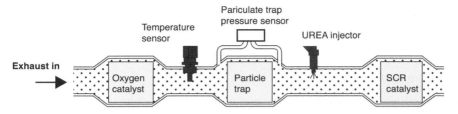

Figure 3.11 Diesel exhaust after treatment system.

Figure 3.11 shows the main components of an exhaust system that incorporates selective catalyst reduction (SCR).

Oxidation catalyst

The oxidation catalyst can operate well in the diesel engine exhaust because there is normally an ample amount of oxygen that results from the engine air supply. During normal operation the catalyst converts carbon monoxide (CO) and hydrocarbons (HC) to carbon dioxide (CO_2), plus some H_2O. In addition, some of the SO_2 is converted into substance that can be collected in the particulate (PM) trap. Sulphur dioxide (SO_2) is recognised as a pollutant because of its role, along with particulate matter, in forming wintertime smog. Studies indicate that SO_2 causes nerve stimulation in the lining of the nose and throat. This can cause irritation, coughing and a feeling of chest tightness, which may cause the airways to narrow. People suffering from asthma are considered to be particularly sensitive to SO_2 concentrations

Diesel particulate filter

Some PM filters are single-use, intended for disposal and replacement once full of accumulated ash. Others are designed to burn off the accumulated particulate either *passively* through the use of a catalyst, or by *active* means such as a fuel burner which heats the filter to soot combustion temperatures. This is accomplished by engine programming to run in a manner that elevates exhaust temperature, in conjunction with an extra fuel injector in the exhaust stream that injects fuel to react with a catalyst element to burn off accumulated soot in the DPF filter, or through other methods. This is known as "filter regeneration". Cleaning is also required as part of periodic maintenance, and it must be done carefully to avoid damaging the filter. Failure of fuel injectors or turbochargers resulting in contamination of the filter with raw diesel or engine oil can also necessitate cleaning. The regeneration process occurs at road speeds higher than can generally be attained on city streets; vehicles driven exclusively at low speeds in urban traffic can require periodic trips at higher speeds to clean out the DPF. If the driver ignores the warning light and waits too long to operate the vehicle above 40 miles per hour, the DPF may not regenerate properly, and continued operation past that point may spoil the DPF completely, so it must be replaced.

Engine performance

Several methods of quoting the power of a vehicle engine have been in use for some time, and they are derived from dynamometer tests under laboratory conditions. Some methods use net power figures and others gross power.

1   **Net power.** Here the engine is tested with all devices such as power steering pump, alternator, water pump, etc., all operating. With this method, the engine is said to be in an *as installed condition*.
2   **Gross power.** The engine is dynamometer tested with no engine driven accessories operating.

Although attempts have been made to set an international standard based on the principle of net power, it is not clear that this has happened, and it is important to understand which method is being used in a particular case because there is likely to be a marked difference between the power output quoted. In the SI system the kilowatt (kW) is the international unit of power. The German national standard (DIN) quotes engine power as PS, which derives from the German word for horse power (pferd-starke). 1 PS is slightly smaller than 1 HP. The original horse power (hp) was derived by James Watt in the 18th century, and it is still widely used.

$$1PS = 0.986HP$$
$$1HP = 0.746kW.$$

Self-assessment questions

3.1   A certain 6 cylinder, 4 stroke diesel engine generates 200 kW at an engine speed of 1200 rev/min and returns a brake specific fuel consumption of 0.28 kg/kWh. Determine the mass of fuel that is injected on each injection period.

Answer 3.1: 0.26 gm

3.2   A turbocharged diesel engine with a compression ratio of 14:1 starts the compression stroke with an air pressure in the cylinder of 2.6 bar (abs) and temperature of 45°C. The compression follows the law $pV^{1.3} = C$.

(a)   Determine the temperature and pressure of the cylinder air at the end of compression.

(b) Describe with the aid of sketch, a device that is used to control boost pressure in a turbocharged engine.

Answer 3.2: 702°K, 80.34 bar

3.3 A traffic pollution survey for Transport for London found that lightly laden heavy diesel vehicles emitted greater amounts of NOx than heavily laden ones.

(a) Explain with reference to air-to-fuel ratios why you think that this could happen.
(b) Why is a broadband oxygen sensor used on diesel engine emission control systems?

3.4 A compression ignition engine with a bore of 120 mm and a stroke of 140 mm has a compression ratio of 18:1.

(a) Determine the size of the clearance volume.
(b) The compression stroke is adiabatic, with $Y = 1.41$, and the pressure at the start of compression is 0.9 bar (abs). Calculate the pressure at the end of compression.
(c) Using the formula for Otto air standard efficiency, determine the air standard efficiency of this engine.

Answer 3.4: (a) 94.1 cm³, (b) 53 bar, (c) 69%

3.5 A single cylinder, 4 stroke diesel engine is tested at constant speed and constant load, and the following results are recorded:
- Bore diameter 210 mm, stroke 350 mm
- Area of indicator diagram 800 mm², base length of diagram 80 mm, indicator spring scale = 1 bar per mm of diagram height
- Engine speed 300 rev/min
- Rope type dynamometer net load 1100 N at a radius of 0.62 m.

Calculate:

(a) the indicated power
(b) the brake power
(c) the mechanical efficiency.

Answer 3.5: (a) 30.31 kW, (b) 21.43 kW, (c) 71%

3.6 A 6 cylinder, 4 stroke diesel engine with a bore of 130 mm and stroke of 160 mm develops a brake power of 350 kW at an engine speed of 1600 rev/min. Calculate:

(a) the brake torque at 1600 rev/min
(b) the BMEP

(c) the indicated power when the mechanical efficiency is 85%
(d) the brake thermal efficiency given that the BSFC is 0.30 kg/kWh and the calorific value of the fuel is 45 MJ/kg.

Answer 3.6: (a) 2088.6 Nm, (b) 20.6 bar, (c) 411.8 kW, (d) 26.7%

3.7   A Morse test on a 5 cylinder compression ignition engine running at 2000 rev/min produced the following results:

Cylinder cut out	None	1	2	3	4	5
Bp.kW	80	63	62	60	61	59
i.p		17	18	20	19	21

Calculate:

(a) the indicated power of the engine
(b) the mechanical efficiency
(c) the amount of friction and pumping losses.

Answer 3.7: (a) 95 kW, (b) 84%, (c) 15 kW

3.8   On a dynamometer test, a 4 cylinder, 4 stroke CIE consumes air at a rate of 3 cubic metres per minute when running at 1800 rev/min. The bore and stroke of the engine are 110 mm and 120 mm respectively. Calculate the volumetric efficiency of the engine at this speed.

Answer 3.8: 59.5%

3.9   An engine on a dynamometer test at a steady speed of 2000 rev/min had an air-to-fuel ratio of 12.4:1 and it consumed fuel at a rate of 0.2 litres per minute.

(a) Given the relative density of the fuel = 0.73, calculate the mass of air used per minute.
(b) The brake thermal efficiency of the engine was 15.3% and the calorific value of the fuel was 47.5 MJ/kg. Calculate the brake power.

Answer 3.9: (a) 1.79 kg, (b) 17 kW

3.10   An atmospherically aspirated 4 cylinder, 4 stroke diesel engine with a bore diameter of 110 mm and a stroke of 120 mm develops a BMEP of 8 bar at an engine speed of 1500 rev/min.

(a) Calculate the brake power of the engine.
(b) The engine is modified to be turbocharged, and this raises the BMEP to 9.5 bar. Compare the power of the modified engine with that of the normally aspirated one.

Answer 3.10: (a) Normal aspiration 45.6 kW, (b) turbocharged 54.1 kW

3.11 An engine operating on the dual combustion cycle has a compression ratio of 16:1 and the maximum pressure reached during the cycle is 65 bar. The temperature at the start of compression is 1 bar and the temperature is 30°C, while the exhaust temperature is 320°C. Determine the temperature at the end of the constant pressure phase of the cycle.

Answer 3.11: 1614°K, 1341°C.

Chapter 4

Transmission

4.1 Lowest and highest gears

The lowest gear ratio is determined by the amount of torque that is required to move the vehicle from rest and overcome gradient resistance, while the highest gear ratio is set by the road speed at which the maximum power that is available to overcome all resistance on a level road is reached. This point is reached where the power-available line intersects the power-required line. The lowest gear will be set to provide the tractive effort required for hill climbing, while the highest gear will be set to provide the maximum speed that will be reached when the power available is equal to the power required.

Example 4.1

A vehicle weighing 1.7 tonnes has an engine that develops a maximum torque of 120 Nm. The rolling radius of the drive axle wheels is 0.30 m. Calculate the gear ratio that will allow the vehicle to ascend a gradient of 25%. Take $g = 9.81$ m/s^2.

$$\textit{The gear ratio to climb gradient} = \frac{\text{drive axle torque}}{\text{engine torque}}$$

$\textit{Drive axle torque} = \text{tractive effort} \times \text{rolling radius of wheel}$

$\textit{Tractive effort} = \text{mass of vehicle} \times \text{gradient}$

When starting from rest, the wind and rolling resistances are negligible.

$\textit{Tractive effort} = 1700 \times 9.81 \times 0.25 = 4170\,\text{N}$

$\textit{Drive axle torque} = 4170 \times 0.30 = 1251\,\text{Nm}$

$\text{Overall gear ratio} = \dfrac{1251}{120} = 10.43.$

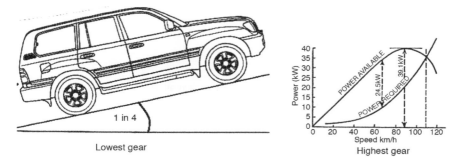

Lowest gear   Highest gear

Figure 4.1 Low and high gears.

Example 4.2

The vehicle in Example 4.1 develops its maximum power at 5800 rev/min, and it is to be designed to have a maximum speed of 160 km/h. Determine the gear ratio required for the vehicle to reach the stated speed.

$$\text{Gear ratio} = \frac{\text{engine speed rpm}}{\text{road wheel speed in rpm}}$$

$$\text{Road wheel speed} = \frac{\text{distance covered in one minute}}{\text{circumference of road wheel}}$$

Rolling radius of road wheel = 0.30 m, circumference $2 \times 3.142 \times 0.3 = 1.89$ m.

$$\textit{Distance covered in one minute} = \frac{160 \times 10^3}{60} = 2670\,\text{m}$$

$$\textit{Road wheel speed}\ \frac{2670}{1.89} = 1413\frac{\text{rev}}{\text{min}}$$

$$\text{Gear ratio} = \frac{5800}{1413} = 4.1:1.$$

Assuming a top gear ratio of 1:1, this figure will represent the final drive ratio.

4.2 The intermediate gear ratios

The concept of ideal tractive effort can be used as a guide to effectiveness of the intermediate gears. The ideal tractive effort assumes that the full rated power of the engine is available across the speed range. A basic principle

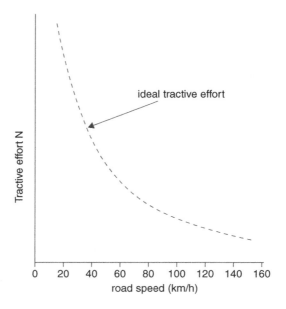

Figure 4.2 Ideal tractive effort

of mechanics is that power $P =$ force $\times$ velocity; in the case of a vehicle, the force propelling the vehicle is the tractive effort Te, and the ideal tractive effort $Te =$ power in Watts $\div$ velocity in $\frac{m}{s}$. If, for a vehicle of given rated power, Te is plotted against velocity, as the curve shown in Figure 4.2 results, this is known as the ideal tractive effort curve.

4.3   Step ratios

An aim of transmission design is to provide a set of gear ratios that produce tractive effort that comes as close as possible to the ideal curve. Figure 4.3 shows tractive effort curves for a four speed gearbox and it is evident that the actual tractive effort does not meet the ideal line all the time. Gear ratios arranged like this are known as step ratios, and they are used in transmission systems of light vehicles.

4.4   Choosing the ratios

In some vehicle transmissions the gear ratios are set to enable the driver of the vehicle to drive at engine speeds between maximum torque and maximum power. The relation between max torque engine speed and max power engine speed is expressed as a ratio that is called the engine speed ratio K =

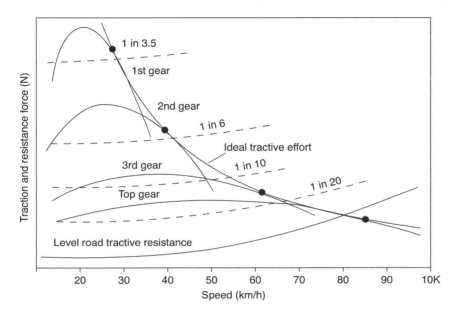

Figure 4.3 Stepped gear ratios

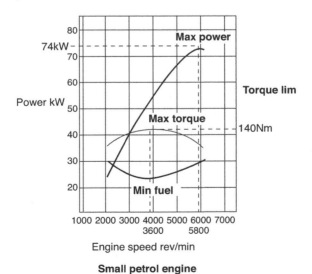

Small petrol engine

Figure 4.4 Engine speed ratio

max torque speed/max power speed. In the case of the engine details shown in Figure 4.4, the max torque occurs at 4000 rev/min and max power occurs at 6000 rev/min. This gives an engine speed ratio of $K = \dfrac{4000}{6000} = 0.67$.

4.5 Graphical method for choosing gear ratios

The example in Figure 4.5 shows the mainshaft output speed of the gearbox plotted against engine speed. In this example, top gear is 1:1, the maximum power is produced at 3200 engine rev/min and the maximum torque is 1800 engine rev/min; the horizontal lines are drawn across as shown to show maximum power and maximum torque. Where the line for top gear crosses the max torque speed line, to change into 3rd gear, the vertical line shows that the mainshaft output speed remains constant. From the point where this vertical line meets the max power line, a line is drawn back to the origin, where this line cuts the maximum torque line another vertical line is drawn and this marks the point where the 2nd gear line is drawn. The process is repeated for 1st gear.

By reading of the graph the various gearbox ratios can be determined, as follows;

top gear = 3200/3200 = 1 to 1
3rd gear = 3200/1800 = 1.77 to 1
2nd gear = 3200/1100 = 2.91 to 1
1st gear = 3200/740 = 4.325 to 1
Engine speed ratio = 1800/3200 = 0.5625

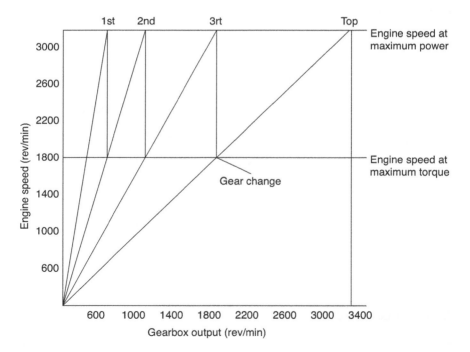

Figure 4.5 Selecting gear ratios

4.6 Some practical gear ratios

Heavy haulage vehicles are normally equipped with a large range of gear ratios some typical examples of which are shown in Table 4.1. Examination of the data in this table will show that there is a numerical relationship between the ratios.

If we take the **Fuller RT-11608** gearbox as an example and calculate the steps between the ratios we should be able to see what the numerical relationship is.

Analysis of the data in Table 4.2 that relates to the **Fuller RT-11608** gearbox shows that there is a common relationship between the gear ratios. The relationship is that of a geometric series. Gear ratios that are arranged in a geometric series produce tractive effort that is very close to the ideal tractive effort.

A geometric progression (GP) is a series where each term is obtained from the preceding term by multiplying or dividing by a constant quantity called the common ratio. The series 1, 2, 4, 8, 16 is an example of a GP where the common ratio is 2. In general a GP can be represented as $a, ar, ar^2, \ldots ar^{n-1}$. The first term is a and the other terms are ar^{n-1} where n is the position of a term in the series.

Example 4.3

A five speed gearbox with a top gear ratio of 1:1 and a lowest gear ratio of 4:1 has the gear ratios that form a GP. Determine the 2nd, 3rd and 4th ratios.

Table 4.1 Heavy vehicle gear ratios

Type of gearbox	Eaton Fuller RT-11608	Eaton Fuller RT/RTO 15615	Eaton Fuller RT-6613	ZF 16S 109
1st gear	10.23	7.83	17.93	11.86
2nd gear	7.23	6.00	14.04	10.07
3rd gear	5.24	4.63	10.96	8.43
4th gear	3.82	3.57	8.61	7.13
5th gear	2.67	2.80	674	5.71
6th gear	1.89	2.19	5.26	4.85
7th gear	1.37	1.68	4.11	3.97
8th gear	1.00	1.30	3.29	3.37
9th gear		1.00	2.61	2.99
10th gear		0.78	2.05	2.54
11th gear			160	2.12
12th gear			125	1.80
13th gear			1.00	1.44

Table 4.2 Analysis of gear ratios

Gear	Ratio	Common ratio K rounded to 1 decimal place
1st	10.23	
2nd	7.23	1st to 2nd $= \dfrac{7.23}{10.23} = 0.7$
3rd	5.24	2nd to 3rd $= \dfrac{5.24}{7.23} = 0.7$
4th	3.82	3rd to 4th $= \dfrac{3.82}{5.24} = 0.7$
5th	2.67	4th to 5th $= \dfrac{2.67}{3.82} = 0.7$
6th	1.89	5th to 6th $= \dfrac{1.89}{2.67} = 0.7$
7th	1.37	6th to 7th $= \dfrac{1.37}{1.89} = 0.7$
8th	1.00	7th to 8th $= \dfrac{1.00}{1.37} = 0.7$

Note: The figures for K in this table are rounded to one decimal place.

Solution 4.3

The ratios form a GP where the first term $a = 4$ and the 5th term $ar^4 = 1$. The first step is to determine the common ratio r.

$$ar^4 = 1,$$

$$\therefore r = \left(\frac{1}{4}\right)^{0.25} = 0.707$$

2nd gear ratio $= ar = 4 \times 0.707 = 2.83$.
3rd gear ratio $= ar^2 = 4 \times 0.707 \times 0.707 = 2.0$.
4th gear ratio $= ar^3 = 1.41$.

4.7   Light vehicle gear ratios

Table 4.4 shows the result of an analysis of the gear ratios that are shown in Table 4.3. It is evident that the steps between the ratios are not equal, which means that the ratios are arrived at by some other means than a geometric series. As shown in Section 4.5, other methods are available for selection of ratios, but it is probable that designers will choose ratios that meet various usage requirements.

Table 4.3 Sample of light vehicle gear ratios

Vehicle	1st	2nd	3rd	4th	5th
Audi A4	3.50	2.12	1.43	1.03	0.84
BMW 320i	4.23	2.52	1.66	1.22	1.00
Chrysler V6	3.31	2.06	1.36	0.97	0.71
Fiat Punto 1	3.91	2.16	1.48	1.12	0.90
Ford Escort	3.42	2.14	1.45	1.03	0.77
Honda	3.25	1.90	1.25	0.91	0.75
Toyota	3.55	1.90	1.31	0.97	0.815
Volvo	3.07	1.77	1.19	0.87	0.70

Table 4.4 Steps between ratios

Vehicle	1st to 2nd	2nd to 3rd	3rd to 4th	4th to 5th
Audi A4	0.61	0.67	0.72	0.82
BMW 320i	0.60	0.66	0.73	0.82
Chrysler V6	0.62	0.66	0.71	0.73
Fiat Punto 1	0.55	0.69	0.80	0.80
Ford Escort	0.63	0.68	0.71	0.75
Honda	0.58	0.66	0.73	0.82
Toyota	0.54	0.69	0.74	0.84
Volvo	0.58	0.67	0.73	0.80

4.8   Increasing the range of ratios

The top of the chart in Figure 4.6 shows the power and torque curves for a large engine that is connected to a gearbox that has two ranges of ratios. When a 16 speed range is used, the engine speed can be kept above about 32 rev/s where high power is available. When an 8 speed range is used, the engine speed can drop to about 26 rev/s where the power is lower but the torque is still high.

4.9   Splitter and range gears

Because of the varying types of operating conditions that heavy vehicles encounter, it is common practice to provide them with a large range of gear ratios. This can be achieved in a number of ways, but the commonest form seems to be to use a conventional four or five speed gearbox and to provide additional ratios by means of an auxiliary gear set built onto the output end of the gearbox. Such gearboxes may be referred to as range or splitter boxes.

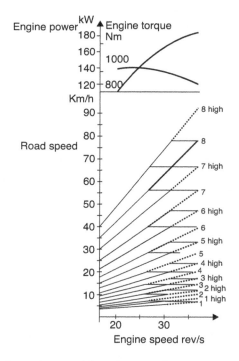

Figure 4.6 Road speed v engine speed for splitter gearbox

In addition heavy vehicle gearboxes employ twin layshafts (countershafts). In this way the large torque from the engine is divided between the layshaft gears, reduces the inter-tooth pressure and permits the use of narrower gears, thus reducing the overall length of the gearbox.

The twin layshaft principle

The auxiliary gearbox is mounted in a separate case on the output end of the main gearbox. It provides two or more ranges of gears so that, in the case of a two speed auxiliary box, the main four speed gearbox can operate as an 8 speed gearbox plus reverse. The auxiliary gearbox is of the constant mesh type, where the range selected is by means of dog clutches which lock the appropriate gears to the auxiliary box main shaft. In the type outlined in Figure 4.7, two countershafts are employed as this divides the torque as in the main gearbox.

The principle of the auxiliary gearbox

To engage the various ratios in the auxiliary gearbox, the dog clutches are moved along the splined mainshaft to lock the appropriate gear wheel to it.

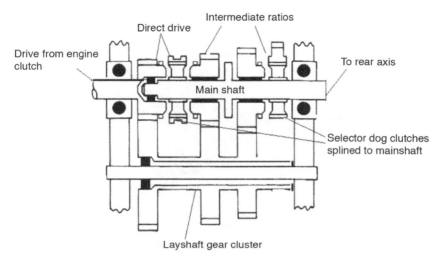

Figure 4.7 A simple auxiliary gearbox

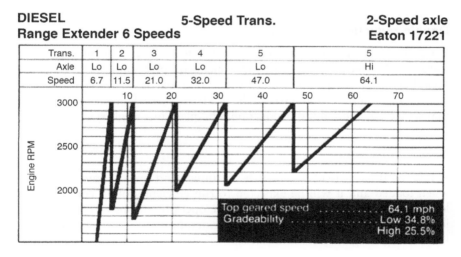

Eaton 2 speed axle

Figure 4.8 Two speed axle performance

4.10   The two speed axle

Figure 4.8 shows that the high ratio of the two speed axle extends the range of ratios available. When the high ratio is selected, it is in effect an extra ratio that acts as an overdrive for higher speed operation. Note that the gradeability of the vehicle is lower when the high range is engaged.

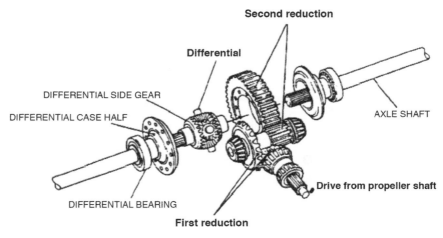

Principle of double reduction final drive

Figure 4.9 A double reduction axle

Double reduction final drive

Large torque multiplication occurs at final drive thus reducing the torque transmitted by the propeller shaft and reducing the probability of transmission "wind up". An example of a double reduction axle is shown in Figure 4.9. The double gear reduction obviates the need for a large diameter crown wheel that would be required if the reduction took place in a single stage.

Gradeability

A lorry without a load may weigh 10 tonnes and up to about 40 tonnes when loaded and it will be expected to run efficiently in all conditions of loading. Although a prime concern of the operator will be to maximise the load at all times, a key feature in this consideration is the performance detail known as gradeability.

Basics of gradeability

Gradeability is the measure of a vehicle's ability to climb a gradient and it is normally expressed as an angle. A fair estimate of gradeability can be obtained as follows:

1   The weight of the vehicle is a force acting vertically downward, by resolving the weight into forces acting parallel and perpendicular to the

surface of the gradient, the force acting parallel to the surface $= mg.\sin\theta$, where m is the mass of the vehicle in kg, g is the gravitational constant and θ is the angle of the gradient in degrees.

2   The opposing force (Te, tractive effort) that propels the vehicle up the gradient is produced by the maximum torque available at the driving axle. This is obtained from the overall gear ratio in the lowest gear, the maximum torque from the engine and the radius of road wheel and tyre.

3   The overall efficiency of the transmission will reduce Te; for the lowest gear, a transmission efficiency of 85% is considered reasonable. The Te will also be affected by the rolling resistance, but at low speed this is disregarded.

By equating the force up the slope to force down the slope (a condition for uniform speed) an expression for the angle θ and hence the angle of the gradient may be determined. This is visualised in Figure 4.10.

Force up slope $Te = (\text{OGR} \times \text{max engine torque} \times \eta)/r$
Force down the slope $= m \times g \times \sin\theta$
$$\sin\theta = \frac{T_e}{mg}$$
The angle θ degrees is used to describe gradeability.

Example 4.4

A vehicle weighing 2 tonnes has a lowest overall gear ratio of 15:1 and the engine develops a maximum torque of 230 Nm. The transmission efficiency is 90% and the road wheels have a rolling radius of 30 cm. Calculate the gradeability of the vehicle.

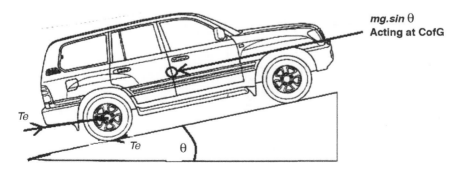

Figure 4.10 Gradeability

Solution 4.4

$$Tractive\ effort = \frac{230 \times 15 \times 0.9}{0.3} = 10350\,\text{N}$$

$$sin\theta = \frac{10350}{2000 \times 9.81} = 0.53$$

$\therefore \theta = 31.8$ degrees.
Gradeability $= 31.8$ degrees.

A limiting factor in gradeability is the coefficient of friction between the tyres and the driving surface

Power and torque transmitted by a friction clutch

Figure 4.11 shows a simple single plate clutch with friction linings that have internal and external radii of r_2 and r_1 respectively.

Let $W =$ spring force pressing the friction surfaces together
$R = mean\ radius\ of\ friction\ plate = \dfrac{(r_1 + r_2)}{2} m.$
$\mu =$ coefficient of friction between friction surfaces
$n =$ number of friction surfaces

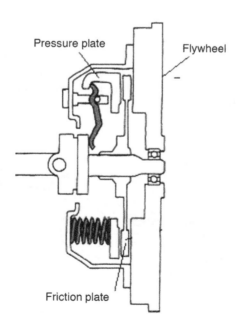

Figure 4.11 A simple friction clutch

Friction force $= \mu W$

Friction torque $T = \mu WR = \mu W \dfrac{(r_1 + r_2)}{2}$

$T = \mu W \dfrac{(r_1 + r_2)}{2}.$

A single plate clutch of the type commonly used in light vehicles has two friction surfaces and torque transmitted, $T = 2\mu W \left(\dfrac{r_1 + r_2}{2} \right)$. For a multiplate clutch with n friction surfaces $T = n\mu W \left(\dfrac{r_1 + r_2}{2} \right)$, where n = number of friction surfaces, μ is the coefficient of friction, W is the spring force, and r_1 and r_2 are the inner and outer radii of the friction surfaces.

Example 4.5

A single plate clutch in a motor car has friction linings with an outer radius of 0.30 m and an inner radius of 0.20 m. The force on the pressure plate is provided by 8 springs each exerting a force of 250 N. The coefficient of friction between the friction surfaces is 0.35. Calculate the maximum torque that can be transmitted by the clutch.

Solution 4.5

$$T = 2\mu W \left(\frac{r_1 + r_2}{2} \right) = 2 \times 0.35 \times 8 \times 250 \left(\frac{0.2 + 0.3}{2} \right) = 350 \, \text{Nm}$$

Example 4.6

A single plate clutch has a friction plate where the mean diameter of the friction linings is 0.075 m with a coefficient of friction 0.4. The total spring force $W = 3$ kN.
   Calculate:

(a) the maximum torque transmitted by the clutch
(b) the power transmitted at an engine speed of 2800 rev/min.

Solution 4.6

(a) $T = 2\mu W \left(\dfrac{r_1 + r_2}{2} \right) = 2 \times 0.4 \times 3000 \times 0.075 = 180 \, \text{Nm}$

(b)   Power $= \dfrac{2\pi TN}{30 \times 1000} = \dfrac{2 \times 3.142 \times 180 \times 2800}{60000} = 52.8 \, \text{kW}$

Torque converter

The three elements of the torque converter (displayed in Figure 4.12) are immersed in transmission fluid that is contained within the space created by the housing of the converter. Rotation of the pumping element causes fluid to pass across to the turbine member, causing it to rotate. From the lower edge of the turbine blades the fluid passes into the stator where the curvature of the blades causes an impulsive force to be generated. The reaction to this force is felt at the turbine blades, resulting in a degree of torque multiplication. The torque multiplication increases until the pump and turbine speeds are the same.

In Figure 4.13, the left-hand scale shows how the torque multiplication is at its highest value when the speed ratio of turbine speed/pump speed is low. At this stage the efficiency of the converter is low because of slip. As the vehicle begins to move, the speed ratio rises with a drop in the torque ratio and a rise in efficiency. When the speed ratio approaches 90%, the

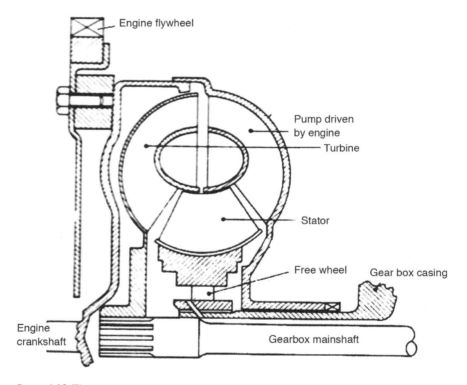

Figure 4.12 The torque converter

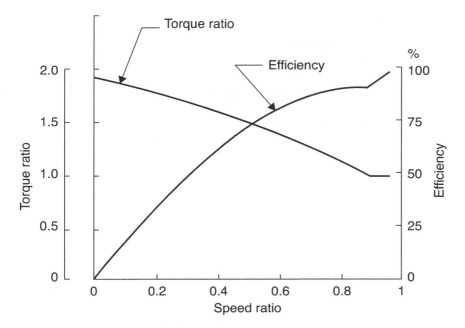

Figure 4.13 Performance characteristics of a torque converter

stator is designed to rotate with other two elements and the converter acts as a hydraulic coupling, and this speed is known as the coupling point. At normal operating speed there is a degree of slip which is a feature of torque converters. The slip leads to excess fuel consumption and emission of CO_2, and this is one of the reasons why alternative types of automatic transmission have come into use.

Dual clutch automatic transmission

The dual clutch gearbox outlined in Figure 4.14 makes use of two friction clutches which are connected to two mainshafts. The mainshaft that operates gears 2, 4 and 6 is hollow and the other mainshaft rotates inside it. The gears are engaged by the normal synchromesh method, where the gears are free to rotate on their respective mainshaft until they are engaged by the synchromesh clutch. The gear selectors and main clutches are electrically operated under the control of the transmission electronic control unit (ECU). A main advantage of the system is the absence slip that is associated with fluid flywheels and torque converters.

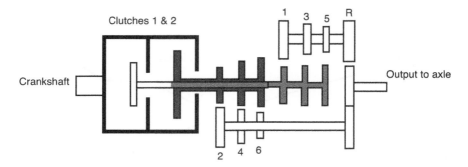

Dual clutch auto transmission

Figure 4.14 A five speed dual clutch transmission

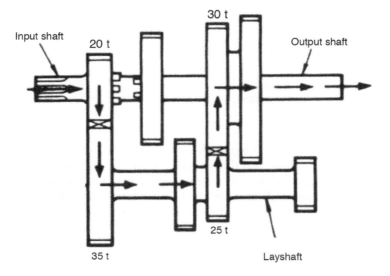

Figure 4.15 Gear ratio

Calculating gear ratios: spur gears

Figure 4.15 shows a simple gear train of the type used in automotive gear-boxes. The gear ratios can be determined as shown in the following example:

$$The\ gear\ ratio = \frac{driven\ gears\ multiplied\ together}{drivers\ multiplied\ together} = \frac{35 \times 30}{20 \times 25} = 2.1:1$$

$$Overall\ transmission\ ratio = gear\ box\ ratio \times final\ drive\ ratio$$

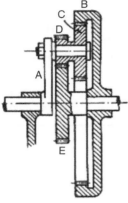

	Revolutions of A	Revolutions of B	Revolutions of C-D	Revolutions of E
(a)	0	−1	$-\dfrac{T_b}{T_c}$	$+\dfrac{T_b}{T_c}\cdot\dfrac{T_d}{T_e}$
(b)	1	1		1
(c)	1	0	$-\dfrac{T_d}{T_e}$	$1+\dfrac{T_b}{T_c}\cdot\dfrac{T_d}{T_e}$

Figure 4.16 Tabular method for epicyclic gear ratios

Epicyclic ratios

The method for determining an epicyclic gear ratio is displayed in Figure 4.16:

1 The carrier A is fixed and prevented from rotating, and the annulus B is rotated by one revolution anti-clockwise. The ratios are shown in line (a) of the table.
2 With all gears locked, the whole train is revolved one revolution clockwise. This is line (b) in the table
3 Line (c) is obtained by adding together corresponding entries in lines (a) and (b), and the result is equivalent to a clockwise rotation of the carrier A through one revolution with the annulus B fixed.

Example 4.8

The following table shows the details of an exercise to determine the gear ratio between input A and output D of an epicyclic gear train. Calculate this ratio.

	Revolutions of A	Revolutions of E	Revolutions of B-C	Revolutions of D
(a) . . .	0	−1	$\dfrac{T_e}{T_c} = \dfrac{21}{30}$	$-\dfrac{T_e}{T_c}\cdot\dfrac{T_b}{T_c} = -\dfrac{21}{30}\cdot\dfrac{27}{24}$
(b) . . .	1	1	1	1
(a) + (b) .	1	0	$1 + \dfrac{21}{30} = \dfrac{17}{10}$	$1 - \dfrac{21}{30}\cdot\dfrac{27}{24} = \dfrac{17}{80}$

Figure Solution 4.8

Solution 4.8

$$\text{Gear ratio} = \frac{revs\ of\ A}{revs\ of\ D} = \frac{1}{17\big/80} = \frac{80}{17} = 4.7:1$$

Self-assessment questions

4.1   A four speed gearbox has a highest ratio 1 and a lowest ratio of 4.6. The gears are arranged in geometric progression. Determine the second and third gear ratios.

Answer 4.1: 2nd gear 2.76, 3rd gear 1.66

4.2   A vehicle weighing 1.5 tonnes has an engine that develops a maximum torque of 120 Nm. The rolling radius of the drive axle wheels is 0.30 m.

(a)   Calculate the gear ratio that will allow the vehicle to ascend a gradient of 25%. Take $g = 9.81$ m/s^2.
(b)   The vehicle engine develops its maximum power at 5800 rev/min, and it is to be designed to have a maximum speed of 160 km/h. Determine the gear ratio required for the vehicle to reach the stated speed.

Answer 4.2: (a) 9.2, (b) 4.1

4.3   A small car weighs 950 kg and the engine produces a maximum power of 30 kW at the road wheels. Rolling and wind resistance are constant at 120 N. If the car starts from rest and climbs a gradient of 1 in 15, calculate:

(a)   the maximum acceleration at a speed of 24 km/h
(b)   the maximum attainable speed of the car.

Answer 4.3: (a) 3.93, (b) 146 km/h

4.4   A vehicle of mass 1400 kg travels at 100 km/h against constant wind and rolling resistance of 800 N. The rolling diameter of the road wheels is 600 mm, and the final drive ratio is 4:1 with a transmission efficiency of 85%. The radius of the road wheels is 0.3 m. Calculate:

(a) the engine power required to propel the vehicle along a level road
(b) the engine power to drive the vehicle up an incline of 1 in 40
(c) the engine torque when climbing the incline.

Answer 4.4: (a) 26.2 kW, (b) 11.43 kW, (c) 101 Nm

4.5   An engine develops brake torque of 125 Nm at 2400 rev/min and transmits the drive through a gearbox where the first motion gears have 24 and 36 teeth respectively. The first (bottom) gear has 45 teeth on the mainshaft and 15 on the layshaft. If the final drive ratio is 3.75:1, calculate:

(a) the speed of the axle shafts in rev/min.
(b) the torque on the propeller shaft.

Answer 4.5: (a) 142 rpm, (b) 562.5 Nm

4.6   A final drive pinion shaft is supported by two bearings 100 mm apart with the pinion outside the bearings. A load of 11 kN acts on the pinion at a point 40 mm from the centre of the nearest bearing, at right angles to the shaft. Calculate the load on each of the bearings.

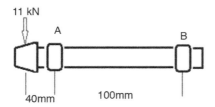

Figure Answer 4.6 Bearing A, 15.4 kN; bearing B, 4.4 kN.

4.7   A single plate clutch has friction plate with a mean radius of 120 mm with a coefficient of friction of 0.4. The clutch is designed to transmit a torque of 180 Nm. Determine the magnitude of the spring force that is required.

Answer 4.7: 597 N

4.8   A certain motorcycle has a clutch with 4 friction plates of mean radius 0.07 m and coefficient of friction 0.25. If the spring force is 800 N, calculate the maximum torque that the clutch can transmit.

Answer 4.8: 703.8 Nm

4.9   An engine develops a torque of 150 Nm at a speed of 3000 rev/min. Using the following data, determine the road wheel speed in rev/min and the torque in:

(a) top gear, which is direct drive
(b) 2nd gear.

First motion shaft pinion: 17 teeth
Constant mesh pinion on the layshaft: 23 teeth
2nd pinion on the layshaft: 18 teeth
2nd speed gear on the mainshaft: 22 teeth
Final drive ratio: 3.7:1
Transmission efficiency: 82%

Answer 4.9: (a) top gear torque 453 Nm, wheel speed 810.8 rpm; (b) 2nd gear torque 750.3 Nm, speed 491.8 rpm

4.10 A certain car engine develops a maximum torque of 140 Nm at a speed of 2500 rev/min. The car gearbox has a 3rd gear ratio of 1.6:1 with an efficiency of 90% and a final drive ratio of 4.2:1 with an efficiency of 87%. The road wheels have a rolling diameter of 0.6 m.

(a) Calculate the road speed of the car when travelling in 3rd gear at maximum torque speed.
(b) If the tractive resistance on the car at this speed is 400 N, calculate the force available to accelerate the car.

Answer 4.10: (a) Speed 42.1 km/h, (b) 2.33 KNm

Chapter 5

Forces and their effects on vehicle performance

5.1 Static reactions at wheels

Figure 5.1 shows a vehicle of mass m kilograms.

The centre of gravity is at a height of h metres above ground level.

The length of the wheelbase $= b$ metres.

The centre of gravity is x metres behind the front axle and y metres in front of the rear axle.

The static reaction forces are R_f at the front axle and R_r at the rear axle.

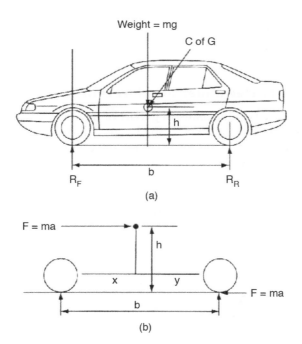

Figure 5.1

Using the principle of moments and taking moments about the point of contact of R_r, we have;

$$R_f b = m.g.y$$

$$R_f = \frac{m.g.y}{b}$$

For R_r take moments about R_f to give:

$$R_r = \frac{m.g.x}{b}$$

Example 5.1

A car weighing 1.2 tonnes has a wheelbase of 2.5 metres, the centre of gravity is 1.2 metres behind the front axle. Determine the static reactions at the front and rear wheels (consult the diagram in Figure 5.2).

$m = 1200$ kg, $g = 9.81$ m/s^2
$b = 2.5$ m, $x = 12$ m, $y = 2.5$ m

$$R_f = \frac{1200 \times 9.81 \times 2.5}{3.7} = 7954\,\text{N} = 7.954\,\text{kN}$$

$$R_r = \frac{1200 \times 9.81 \times 1.2}{3.7} = 3.818\,\text{kN}$$

5.2   Load transfer under acceleration

The inertia force acts at the centre of gravity of the vehicle, which is h metres above the road level. If the acceleration is a.m/s^2 and the mass of the vehicle is m.kg, the inertia force acting at the centre of gravity is F = m.a. This force acting at h metres above road level exerts a turning effect about the point of

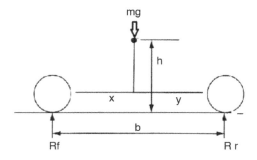

Figure 5.2

contact between the rear wheels and the road which effectively reduces the load on the front axle and increases it on the rear axle.

Maximum acceleration

The limiting factor is the frictional grip between the tyre and the driving surface. When the friction force is equal to the accelerating force, wheel slip will occur. This is the factor that determines the maximum acceleration of the vehicle.

Example 5.2

A small saloon car has a mass (weight) of 1.5 tonnes. The centre of gravity is 0.7 m above road level and it is placed 1 m behind the front axle and 1.6 m in front of the rear axle. The coefficient of friction between the tyres and the road surface is 0.50 (take $g = 9.81$ m/s^2). Determine the maximum acceleration that can be achieved:

(a) when the drive is through the front wheels
(b) when the drive is through the rear wheels
(c) when the drive is through all four wheels with the 3rd differential locked.

Under acceleration the load transfer due to force acting at the centre of gravity, take moments about the point of contact of the rear wheel and the road (refer to Figure 5.3):

$$CWM = ACWM$$
$$R_f b + F.h = m.g.y$$
$$R_f = \frac{m.g.y}{b} - \frac{F.h}{b}$$

$\dfrac{m.g.y}{b}$ is the static reaction at the rear wheels

$\dfrac{F.h}{b}$ is the load transfer

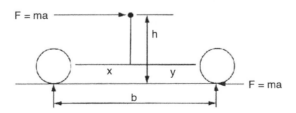

Figure 5.3

Solution 5.2(a) front wheel drive

The tractive effort (Te) is the force that is applied at the point of contact between the tyres and the road, and it is equal and opposite to the inertia force that acts at the centre of gravity of the vehicle. It is dependent on friction between tyres and the road and the force pressing down, as the force pressing down is the static reaction minus load transfer, the tractive effort.

$$T_e = \mu\left(R_f - \frac{F.h}{b}\right).$$

The tractive effort Te is equal and opposite to the inertia force F that acts at the centre of gravity of the vehicle. If the maximum possible acceleration is a m/s^2, the inertia force $F = m.a$ and this is the F in the load transfer term, $\frac{F.h}{b}$.

We may now say that $T_e = m.a = \mu\left(R_f - \frac{F.h}{b}\right)$, and by writing $m.a$ in place of F inside the brackets and removing them, we have:

$$ma = \mu R_f - \frac{\mu m a h}{b}. \qquad\qquad 5.5$$

The numerical values required to evaluate the acceleration a are:

$m = 1500$ kg, $h = 0.70$ m, $x = 1.0$ m, $y = 1.6$ m, $b = 2.6$ m, $\mu = 0.50$.

$R_f = \dfrac{mgy}{b} = \dfrac{1500 \times 9.81 \times 1.6}{2.6} = 9055\,N$. Substituting these values in

equation 5.5 gives; $1500\,a = 0.50 \times 9055 - \dfrac{0.50 \times 1500 \times a \times 0.70}{2.6}$.

$1500a = 4527.5 - 202\,a$

$1702\,a = 4527.5$

$$a = \frac{4527.5}{1702} = 2.66\,\frac{m}{s^2}$$

Solution 5.2(b) rear wheel drive

Under acceleration the load transfer is to the rear axle and the tractive effort $T_e = \mu\left(R_r + \dfrac{m.a.h}{b}\right)$.

$$R_r = \frac{m.g.x}{b} = \frac{1500 \times 9.81 \times 0.7}{2.6} = 3962\,N$$

$$T_e = 1500a = \mu\left(3962 + \frac{1500 \times 0.7.a}{2.6}\right).$$

$$1500a = 0.5 \times 3962 + 0.5\left(\frac{1500 \times 0.7\,a}{2.6}\right)$$

$$1500a = 1980 + 202\,a$$

$$1298a = 1980.$$

$$a = \frac{1980}{1298} = 1.53\,\text{m}/\text{s}^2.$$

Solution 5.2(c) four wheel drive – 3rd differential locked

In four wheel drive, the force pressing the tyres to the road is the weight of the vehicle.

$$T_e = m.a = \mu.m.g$$

$$m.a = \mu mg$$

$$a = \mu g$$

$$a = 0.50 \times 9.81 = 4.905\,\text{m}/\text{s}^2$$

5.4 This example demonstrates that spreading the tractive effort across the four wheels leads to a large increase in the maximum acceleration. This improved traction is useful in wet and icy conditions and improves the starting off and hill climbing ability of a vehicle. However, because the front and rear axle perform differently under certain conditions, such as cornering, there is a problem known as "wind up" in the transmission, and to overcome the problem, a differential gear is fitted between the front and rear axles. This additional differential, which is known as the 3rd differential, balances the torque between the two axles, which means that loss of grip in one axle leads to loss of grip in both. In order to deal with this loss of grip, the 3rd differential is provided with a lock which can be deployed as required.

The differential lock can be situated at the front drive clutch, where it locks the front drive shaft to the differential carrier. Another form of differential lock uses the ABS braking system to lock any wheel that may be losing grip.

5.3   Electronically controlled four wheel drive

The Haldex coupling incorporates a hydraulically operated multiplate clutch which is placed in the driveline between the front and rear axles. It is controlled electronically via the CAN bus in association with other vehicle systems. An outline of the modes of operation is given in the table in Figure 5.5.

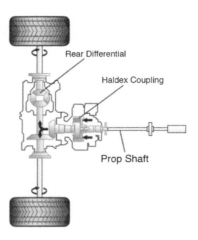

Figure 5.4 An electronically controlled four wheel drive system.

	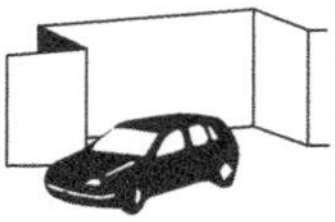 Parking	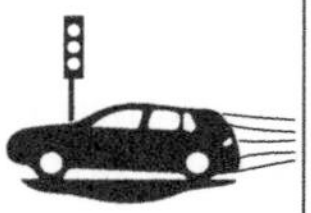 Acceleration	 High-speed driving
Difference in speed beween front and rear axles	low	high	low
Torque required at rear axle	low	high	low
Condition of multi-plate clutch	low contact pressure	high contact pressure, up to maximum, EDL control system can increase contact pressure	closed, as required
Input signals	- engine torque - engine speed - accelerator pedal position - four-wheel sensors	- engine torque - engine speed - accelerator pedal position - four-wheel sensors	- engine torque - engine speed - accelerator pedal position - four-wheel sensors

Figure 5.5 Operating details of Haldex type four wheel drive

5.4   Overturning effects

When a vehicle is being steered around a curve it is subjected to centrifugal force which acts at the centre of gravity of the vehicle. The centrifugal force that is created attempts to overturn the vehicle and this action is countered by the turning effect of the weight acting downward through the centre of gravity. Modern motor cars are designed so that normal cornering will not overturn the vehicle; however, should the outside wheel bump up against a kerb, the shock effect may cause the vehicle to overturn.

With reference to Figure 5.6 and taking moments about R_B;

clockwise moments = anti clockwise moments

$$\frac{m.g.d}{2} = \frac{m.v^2.h}{r}$$

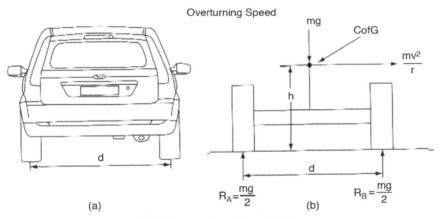

(a)   (b)

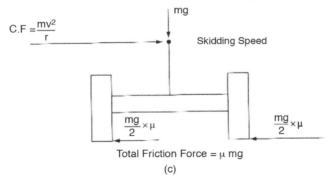

(c)

Figure 5.6 Vehicle cornering on a level track

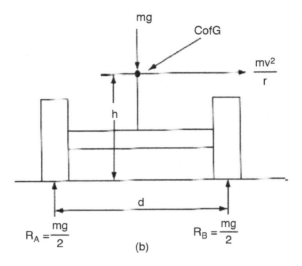

Figure 5.6b Overturning effect

Mass of vehicle m kg
Gravitational constant g
Track width d metres
Height of centre of gravity h metres
Radius of turn r metres
Velocity of vehicle v metres/second.

When the overturning moment is equal to the righting moment, the wheels on the inside of the corner are lifting from the road and the vehicle is about to overturn.

$$\frac{mv^2h}{r} = \frac{mgd}{2}$$
$$v^2 = \frac{grd}{2h}.$$
$$v = \sqrt{\frac{grd}{2h}}$$

Example 5.3

A vehicle has a track width of 1.8 m and its centre of gravity is 0.55 m above ground level. What is the maximum speed that it can travel round a bend of 30 m radius?

Solution 5.3

Overturning velocity $v = \sqrt{\dfrac{grd}{2h}}$

$$v = \sqrt{\dfrac{9.81 \times 30 \times 1.8}{2 \times 0.55}}$$

$$v = \sqrt{481.6}$$

$$v = 21.95\, m/s = 79\, km/h$$

5.5   Overturning on a curved banked track

The main forces acting at the centre of gravity of the vehicle are;

- Centrifugal force acting horizontally at the centre of gravity ($\dfrac{mv^2}{r}$), where m = mass of vehicle in kg, v = velocity of vehicle in metres per second and r = radius of the corner.
- Weight mg Newtons acting vertically downward through the centre of gravity of the vehicle.

The other dimensions are the track width of the vehicle d metres, the height of the centre of gravity h metres and the angle of the banking is θ degrees.

For this purpose we are concerned with forces that are acting parallel to and perpendicular to the surface of the slope. The forces are the components of the two main forces and these are;

- The component of the centrifugal force parallel to the slope which is $\dfrac{mv^2}{r}\, cos\,\theta\, Newtons$, the component of the weight parallel to the slope which is $mg\, sin\,\theta\, Newtons$ acting in the opposite direction.
- The components of the forces acting perpendicular to the slope are $m.g\, cos\theta$ Newtons and $\dfrac{mv^2}{r}\theta\, Newtons$.

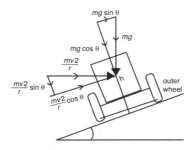

Figure 5.7 Vehicle on curved and banked track

Overturning effect

Taking moments about the outside wheel;

The overturning moment $= \dfrac{mv^2}{r}\cos\theta.h.$

The righting moments are;

$$\frac{mv^2}{r}\sin\theta\frac{d}{2} + mg\cos\theta\frac{d}{2} + mg\sin\theta h$$

At the instant of overturning the overturning moment is equal to the righting moment as follows;

Overturning moment $=$ Righting moment

$$\frac{mv^2}{r}\cos\theta.h = \frac{mv^2}{r}\sin\theta\frac{d}{2} + mg\cos\theta\frac{d}{2} + mg\sin\theta h$$

Simplify by dividing both sides by $\cos\theta$ gives,

$$\frac{mv^2}{r}.h = \frac{mv^2}{r}\tan\theta.\frac{d}{2} + mg.\frac{d}{2} + mg\tan\theta h$$

Dividing through by m and h gives:

$$\frac{v^2}{r} = \frac{v^2}{r}\tan\theta.\frac{d}{2.h} + g.\frac{d}{2.h} + g\tan\theta$$

Grouping the terms containing $\dfrac{v^2}{r}$ gives:

$$\frac{v^2}{r} - \frac{v^2}{r}\tan\theta.\frac{d}{2.h} = g.\frac{d}{2.h} + g\tan\theta$$

$$\frac{v^2}{r}\left(h - \tan\theta.\frac{d}{2}\right) = g\left(\frac{d}{2} + h\tan\theta\right)$$

$$v = \sqrt{gr\left(\frac{\left(\frac{d}{2} + h.\tan\theta\right)}{\left(h - \frac{d}{2}\tan\theta\right)}\right)}$$

Example 5.4

A car travels on a curved banked road of 45 m radius which is banked at an angle of 20 degrees. The track width of the car is 1.45 metres and its centre of gravity is 0.7 metres above ground level. Take $g = 9.81$ m/s^2.

Calculate the overturning speed of the car.

$$v = \sqrt{gr\left(\frac{\left(\frac{d}{2}+h.\tan\theta\right)}{\left(h-\frac{d}{2}\tan\theta\right)}\right)}\; m/s.$$

$\tan 20° = 0.364$

$$v = \sqrt{9.81\times45\left(\frac{0.725+0.7\times0.364}{0.7-0.725\times0.364}\right)}$$

$$v = \sqrt{441.5\times\frac{0.988}{0.436}}$$

$$v = \sqrt{992.4} = 31.5\,m/s = 113.4\,km/h$$

5.6   Skidding on curved banked track

Refer to Figure 5.7. The force attempting to push the vehicle up the slope:

$$\frac{mv^2}{r}\cos\theta - mg\sin\theta.$$

Force opposing $\mu mg\cos\theta + \mu\dfrac{v^2}{r}\sin\theta$

Equating these because when they are equal skidding is about to occur.

$$\frac{mv^2}{r}\cos\theta - mg\sin\theta = \mu mg\cos\theta + \mu\frac{mv^2}{r}\sin\theta$$

Dividing through by cos θ and m

$$\frac{v^2}{r} - g\tan\theta = \mu g + \mu\frac{v^2}{r}\tan\theta$$

$$\frac{v^2}{r}(1-\mu\tan\theta) = g(\mu+\tan\theta)$$

$$v = \sqrt{gr\left(\frac{\mu+\tan\theta}{1-\mu\tan\theta}\right)}$$

Example 5.5

A car travels round a curved banked track that has a radius of 35 m and a banked angle of 20 degrees, the coefficient of friction between the tyres and road is 0.7. Determine the velocity at which skidding will occur. Take $g = 9.81$ m/s².

Solution 5.5

$$v = \sqrt{9.81 \times 35 \left(\frac{0.7 + 0.364}{1 - 0.7 \times 0.364} \right)} = \sqrt{343.5 \times 1.42} = 22.1 m/s$$

5.7 Electronic stability program

The earlier examples show that when a vehicle is negotiating a curve, the wheels on the inside of the curve bear a decreasing proportion of the vehicle weight. At some point between zero speed and skidding or overturning speed, the friction force may not be sufficient to support the tractive effort and the inner wheel may start to spin. This may affect the directional stability of the vehicle and require remedial action. The electronic stability program is designed to use the traction control system to apply the brake of the spinning wheel, and if necessary control engine power, so that the vehicle does not veer from the path required by the driver. Some vehicles that are fitted with an active steering system allow the ESP (electronic stability program) to operate the steering for which purpose electrically operated steering systems have been designed.

5.8 Effect of powertrain inertia

Flywheels

The function of a flywheel is to carry the engine over the strokes when no energy is being produced and to ensure that the engine speed does not vary too greatly between one firing stroke and the next. In order to perform these functions, the flywheel has a mass of several kilograms, which gives it considerable inertia, depending on the size of the engine. A property of the flywheel that enables it to perform its function is the *moment of inertia*.

Moment of inertia

The moment of inertia of a mass m kg that is constrained to move on a circular path is a product of the mass and the square of the distance from the centre of rotation. This distance is known as the **radius of gyration** and it is normally represented by the letter k.

The moment of inertia is denoted by I; $I = mk^2$, where m is in kg and k is in metres.

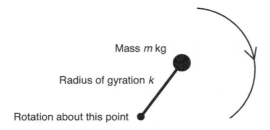

Figure 5.8 Moment of inertia

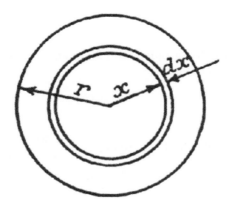

Figure 5.9 Moment of inertia of a simple disc

When a body is rotating, the whole of its mass may be considered to be placed in a ring at a radius. That is the radius of gyration, as denoted by the letter k in Figure 5.8.

5.9 Moment of inertia of a uniform disc

The moment of inertia of the whole disc is the sum of the moments of inertia of all the rings of width dx from the centre of the disc to the outer edge (as displayed in Figure 5.9).

This sum is denoted by $\int dm.r^2$.
Let m kg be the total mass of the disc so that unit area will have a mass of $\dfrac{m}{\pi r^2}$.

Mass of the elementary ring of width $dx = \dfrac{m}{\pi r2}.2\pi x\,dx = \dfrac{2m}{r^2}x\,dx$

Moment of inertia of each ring $= \dfrac{2m}{r^2}.x\,dx.x^2 = \dfrac{2m}{r^2}x^3\,dx$

Moment of inertia of the disc $I = \dfrac{2m}{r^2}\displaystyle\int_0^r x^3\,dx = \dfrac{2m}{r^2}\left[\dfrac{x}{4}\right]_0^r$

When the square bracket is evaluated, $I = \dfrac{mr^2}{2}$

The radius of gyration is denoted by the letter k, where $k = \dfrac{r}{\sqrt[2]{2}}$

The moment of inertia of a simple disc $I = mk^2$ where k is the radius of gyration in metres and $m = $ mass in kg.

Example 5.6

A simple disc flywheel has a radius of 0.30 m and a mass of 50 kg. Determine its moment of inertia.

Solution 5.6

$I = mk^2$
The radius of the flywheel $= 0.30$ m $\therefore k = \dfrac{0.30}{\sqrt[2]{2}} = \dfrac{0.30}{1.414} = 0.212\,\text{m}$

Moment of inertia $I = 50 \times 0.212^2 \text{ kg m}^2 = 2.25 \text{ kg m}^2$.

5.10   Kinetic energy of rotation

A small mass dm rotates at an angular velocity of ω radians per second to give a linear velocity of $v = \omega r$ (displayed in Figure 5.10).

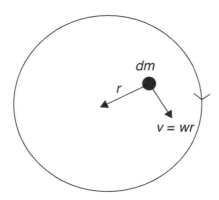

Figure 5.10 Rotating mass

The kinetic energy of the small object of mass $= dm.\dfrac{v^2}{2}$

$= dm.\dfrac{(\omega r)^2}{2}$

The total KE of the disc $= \displaystyle\int dm.\dfrac{(\omega r)^2}{2}$

But $\displaystyle\int dm.r^2 = I$

$\therefore$ Total KE of disc $= \dfrac{1}{2}I\omega^2$

Example 5.7

Calculate the kinetic energy stored in a simple disc flywheel of radius 0.30 m and mass 10 kg when it is rotating at 3000 rev/min.

Solution 5.7

KE of disc $= \dfrac{1}{2}I\omega^2.$

$I = mk^2 = 10\times\dfrac{0.3\times0.3}{1.414\times1.414} = 0.45\,\text{kgm}^2$

angular velocity $\omega = \dfrac{3000\times2\pi}{60} = 314.2\,\text{radians per second}$

$KE = \dfrac{0.45\times314.2\times314.2}{2} = 22212\,J = 22.2\,kJ$

Torque and angular acceleration

With reference to Figure 5.11, let T be the torque required to the disc an angular acceleration of α radians/s^2.

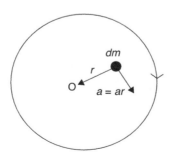

Figure 5.11 Torque and angular acceleration

The **linear** acceleration of small mass $dm = \alpha r$
Force applied to $dm = dm\alpha r$ (mass × acceleration)
Torque required to accelerate $dm = dm.\alpha\, r.r$
$T = dm.\alpha\, r^2$.

Torque T required to accelerate the whole disc $= \int dm.r^2.\alpha$

But $\int dm.r^2 = I$

$\therefore T = I\alpha$.

Example 5.8

A flywheel has a mass of 7 kg and a radius of gyration k of 0.20 m. Determine the angular acceleration when a torque of 20 Nm is applied to the flywheel.

Solution

First find I from $I = mk^2$

$$I = 7\times0.20\times0.20 = 0.28\,\text{kgm}^2$$

Then use $T = I\alpha$ to find α.

$$\alpha = \frac{T}{I} = \frac{20}{0.28} = 71.43\,\frac{rad}{s^2}$$

5.11   Inertia of the entire powertrain

The following examples show how powertrain inertia affects vehicle performance.

Example 5.9

A small four wheel vehicle with a weight of 910 kg has an overall gear ratio of 14:1. The wheels have a rolling radius of 30 cm and the total moment of inertia of the wheels and transmission amounts to 8.5 kg m². The engine moment of inertia is 0.85 kg m² and the torque produced is 140 Nm. In the gear ratio in question the transmission efficiency is 90%, rolling resistance and other effects amount to a constant force opposing motion of 450 N. Determine the maximum acceleration of the vehicle on an upward gradient of 1 in 20.

Solution 5.9

Let the linear acceleration be A m/s^2.
The angular acceleration of the road wheels $\alpha = \dfrac{A}{r}$ radians per second², where r = radius of wheel.
The radius of the wheels = 0.30 m

Angular acceleration of wheels $\alpha = \dfrac{A}{r} = 3.33A$ rad/s^2.

Angular acceleration of engine $= 14 \times 3.33A = 46.62A\dfrac{rad}{s^2}$.

Force at wheels on the slope $= R + \dfrac{910 \times 9.81}{20} + 910A\ 445 + 446.3 +$

$910A = 891 + 910A$.

Torque at wheels for this force $T = (891 + 910A)0.3 = (267.3 + 273A)Nm$

Inertia torque of wheels $= I\dfrac{A}{r} = 8.5 \times 3.33A = 28.3ANm$

Total torque at wheels $= 267.3 + 273A + 28.3A = (267.3 + 301.3A)Nm$.

Torque that the engine transmits to the wheels $= \left\{ \dfrac{267.3 + 301.3A}{G} \right\} \dfrac{100}{90}$.

Inertia torque of the engine $= IG\dfrac{A}{r} = 0.85 \times 3.33A \times G = 2.83AG$ Nm

Torque required from engine $140 = \left\{ \dfrac{267.3 + 301.3A}{G} \right\} \dfrac{100}{90}. + 2.83AG$

$0.9G(140 - 2.83AG) = 267.3 + 301.3A$

G the overall gear ratio $= 14$

$\therefore 1764 - 499A = 267.3 + 301.3A$

$\Lambda\ 800A = 1496.7$

Acceleration A $= \dfrac{1496.7}{800} = 1.87$m/second^2

5.12   Other forces acting on a moving vehicle

Aerodynamic resistance

As a vehicle moves through the air, it displaces the air in front of it, and this causes a build-up or pressure. The force on the front of the vehicle that results from the pressure build-up is dependent on several factors, including the frontal area of the vehicle. The force that is created is also known as drag, and it consists primarily of three elements:

1   Frontal pressure, or the effect created by a vehicle body pushing air out of the way.
2   Rear vacuum, or the effect created by air not being able to fill the hole left by the vehicle body.
3   Boundary layer, or the effect of friction created by slow-moving air at the surface of the vehicle body.

Aerodynamic Resistance (AR):

$$AR = \dfrac{1}{2}\rho A_f C_d \left(\dfrac{v}{3.6} \right)^2$$

$$AR = \frac{1}{2}\rho A_f C_d \left(\frac{v_{car} \pm v_{wind}}{3.6} \right)^2$$

where:

AR = air resistance [N]

ρ = air density [kg/m³] $\approx$ 1.202 kg/m³, at sea level and at 15 C

A_f = car frontal area [m²] $\approx$ 1.2: 3.2 m², for small and mid-size cars

C_d = coefficient of aerodynamics resistance (drag coefficient) $\approx$ 0.2: 0.5 for c

v = car relative velocity [km/h]

v_{car} = car velocity [km/h] ($v_{car} = v$, at stand still wind, $v_{wind} = 0$)

v_{wind} = wind velocity [km/h]

Example 5.11

A car with a frontal area of 2.5 m² and Cd of 0.30 travels at 50 km/h on a level road and windless day. Calculate the aerodynamic resistance that the car creates.

Solution 5.11

$$AR = \frac{1}{2}\rho A_f C_d \left(\frac{v}{3.6} \right)^2$$

$$AR = \frac{1}{2}\rho A_f C_d \left(\frac{v_{car} \pm v_{wind}}{3.6} \right)^2$$

$$AR = \frac{1}{2} \times 1.202 \times 2.5 \times 0.30 \left(\frac{50}{3.6} \right)^2 = 87\,N$$

Example 5.12

Calculate the aerodynamic resistance of the car in Example 5.11 if the speed is increased to 100 km/h against a head wind of 15 km/h.

Solution 5.12

$$AR = \frac{\rho A C_d}{2} \frac{\left(v_{car} + v_{wind} \right)^2}{3.6}$$

$$AR = \frac{1.202 \times 2.5 \times 0.30 \left(\frac{115}{3.6} \right)^2}{2}$$

$$AR = 0.451 \times 1020.5 = 500 N$$

Rolling resistance

Rolling resistance is caused by flexing of the tyre, particularly in the tread area, and it is affected by the composition of the rubber. Rolling resistance $F_r = k_R \times F_w$. where k_R is the rolling resistance factor and F_w is the force

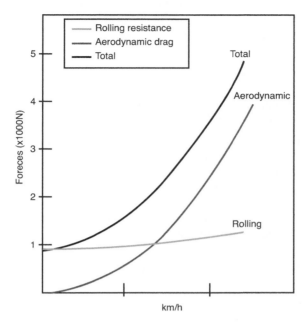

Figure 5.12 Approximate effect of aerodynamic drag and rolling resistance

pressing the tyre to the road. The rolling resistance factor is quite small, varying from approximately 0.01 to 0.02, according to tyre pressure and type of rubber used; it remains constant up to approximately 120 km/h after which it increases sharply (Figure 5.12).

Self-assessment questions

5.1 A vehicle of mass 1.8 tonnes has a wheel base of 2.6 m. The centre of gravity is 0.75 m above ground level and midway between the axles. If the coefficient of friction between tyres and road is 0.6, calculate the maximum acceleration when the drive is by:

(a) front axle
(b) rear axle
(c) front and rear (4 wheel drive, no 3rd differential).

Answer 5.1: (a) 2.51 m/s^2, (b) 3.56 m/s^2, (c) 5.89 m/s^2

5.2 The track of a car is 1.4 m and the centre of gravity is in the centre of the car at a height of 0.65 m above ground level.

(a) If the coefficient of friction between tyre and road is 0.5, determine the greatest speed in km/h that the car can take a corner of radius 20 m without skidding occurring.

(b) Assuming that friction is adequate to prevent skidding, determine the greatest speed that the vehicle can negotiate the 20 m radius corner without overturning.

Answer 5.2

(a) 35.6 km/h, (b) 52.3 km/h we

5.3   A four wheel vehicle has a track width of 1.4 m and the centre of gravity is 0.80 m above ground level.

(a) Given that the coefficient of friction between tyres and road is 0.5, calculate the skidding speed in km/h at which the vehicle can take a corner of 60 m radius on a track banked at 20 degrees.

(b) Calculate the overturning speed for the vehicle on the same banked track.

(c) Write a short description of the way in which an ESP interacts with the engine and braking controls to help a driver when a wheel starts to lose traction.

Answer 5.3: (a) 58.3 km/h, (b) 118 km/h

5.4   A four cylinder engine has a moment of inertia 0.4 kg m^2.

(a) Calculate the engine torque required to accelerate the engine from 3000 rev/min to 5000 rev/min in a time of 1.2 seconds.

(b) As a result of some tuning work, a lighter flywheel and pistons are fitted and the moment of inertia of the engine is reduced by 0.30 kg m^2. Calculate the engine torque required to accelerate the engine across the same speed range. What difference does the change make to the amount of engine torque that is available to accelerate the vehicle?

Answer 5.4: (a) 69.8 Nm, (b) 52.38 Nm, extra 17.46 Nm

5.5   A four wheel vehicle weighing 1.5 tonnes has its centre of gravity at a height of 0.6 metres above ground level, midway between the front and rear axles. The track width is 1.4 m. Calculate the force acting at the inner wheels when the vehicle is taking a corner of 68 m radius, on a level road, at a speed of 64 km/hour.

Answer 5.5: 4362.5 N

5.6   A four wheel vehicle with a weight of 1100 kg has an overall gear ratio of 14:1. The wheels have a rolling radius of 30 cm and the total moment of inertia of the wheels and transmission amounts to 8.5 kg m^2. The engine moment of inertia is 0.85 kg m^2 and the torque produced is 140 Nm. In the gear ratio in question the transmission efficiency is 90%, rolling resistance and other effects amount to a constant force opposing motion of 450 N. Determine the maximum acceleration of the vehicle on an upward gradient of 1 in 20.

Answer 5.6: 1.75 m/s^2

Chapter 6

Simple harmonic motion and vibrations

6.1   Simple harmonic motion (SHM)

In order to show that a motion is simple harmonic the following must occur (1) its acceleration must be directly proportional to its displacement from the fixed point in its path (2) the acceleration must always be directed towards the fixed point.

The traditional approach to a study of simple harmonic motion is to consider a point P that is constrained to move in a circular path about a point O.

With reference to Figure 6.1, let point P move along a circular path with constant velocity v, and let N be the projection of P on to the diameter AB.

The velocity of N is equal to the horizontal component of v.

$$
\begin{aligned}
\text{i.e. velocity of } N &= v \sin \theta, \text{ since } v = \omega R \\
&= \omega R \sin \theta \\
&= \omega NP \\
&= \omega \sqrt{\left(OP^2 - ON^2 \right)} \\
&= \omega \sqrt{\left(R^2 - x^2 \right)}
\end{aligned}
$$

The maximum velocity of $N = v = \omega R$. That is when $\theta = 90°$ or $270°$.

Because the point P moves along a circular path about centre O, it has a centripetal acceleration of $\dfrac{v^2}{R}$ and the only acceleration that N may have is the horizontal component of this centripetal acceleration and that is;

$$
\text{acceleration} = \frac{v^2}{R} \cos \theta.
$$

$$
\begin{aligned}
\text{The acceleration of } N &= \omega^2 R \cos \theta \\
&= \omega^2 x.
\end{aligned}
\tag{6.1}
$$

The maximum acceleration of $N = \omega^2 R$ m/s^2 occurs when $\theta = 0°$ or $180°$, that is when $\cos \theta = 1$.

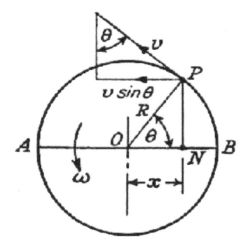

Figure 6.1 Simple harmonic motion SHM

Equation 6.1 shows that the acceleration of N is proportional to the displacement x and is always directed towards O, so that the motion of N is simple harmonic.

The **amplitude** of the vibration is the maximum displacement of N from mid-position O, and this is equal to the radius of the circle R.

The **periodic time** t is the time required for one N to complete two strokes, and this is the time that P takes to make one revolution.

$$\text{Thus the periodic time } t = \frac{distance\ covered}{velocity}$$

$$= \frac{\text{circumference}}{\text{velocity}}$$

$$= \frac{2\pi R}{\omega R}$$

because velocity $v = \omega R$

Periodic time $t = \dfrac{2\pi}{\omega}$ (6.2)

By using equation 6.1, acceleration a $= \omega^2 x$, where $x =$ displacement, it can be shown that $\omega = \sqrt{\dfrac{\text{acceleration}}{\text{displacement}}}$

For simple harmonic motion:

$$\text{Periodic time } t = 2\pi\sqrt{\frac{displacement}{\text{acceleration}}} \qquad (6.3)$$

$$\text{Frequency of vibration } f = \frac{1}{t} \; vibrations \; per \; second \qquad (6.4)$$

Example 6.1

A body moving with S.H.M has an amplitude of 1.3 m and a periodic time of 3 seconds. Calculate (a) the maximum velocity, (b) the maximum acceleration and (c) the frequency.

Solution 6.1

(a) $t = \dfrac{2\pi}{\omega}$. from equation 6.2

$$\therefore \omega = \frac{2\pi}{t}$$
$$= 2\pi/3 \text{ radians/second}$$

Max velocity $= \omega R = (2\pi/3) \times 1.3 = 2.72$ m/s.

(b)   maximum acceleration $= \omega 2R = (4\,\pi 2\,/\,9) \times 1.3 = 5.7$ m/s².

(c)   frequency $= \dfrac{1}{t} = \dfrac{1}{3} = 0.33$ Hz

6.2   Vibration of a helical coil spring

The coil spring in Figure 6.2 is fixed at the top end and at the bottom end is suspended a weight of W kg $= W$ g Newtons. This weight is free to move in a vertical plane relative to the datum position O.

If the weight is pulled down by a distance x and then released, the weight will vibrate (oscillate) to and fro about the datum line O.

If the stiffness of the spring is S Newtons per metre, the restoring force exerted by the spring $= Sx$.

$Sx = Wa$, where W is the mass in kg and $a = $ acceleration in m/s²

$\therefore a = \dfrac{Sx}{W}$, i.e. acceleration is proportional to displacement.

This is the condition for the motion to be SHM.

The periodic time $t = 2\pi \sqrt{(\text{displacement/acceleration})}$

$\therefore t = 2\pi \sqrt{(W/S)}$

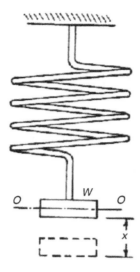

Figure 6.2 SHM of open coiled spring

Because S = W*g*, where *d* = static deflection of the spring and *g* is the gravitational constant, the periodic time may be written as;

$$t = 2\pi\sqrt{\frac{d}{g}}$$

Example 6.2

A suspension coil spring has a static deflection of 250 mm. Calculate the frequency of vibration. Take *g* = 9.81 m/s².

Solution 6.2

$$t = 2\pi \sqrt{(d/g)},$$
$$\therefore t = 2\pi \sqrt{(0.25/9.81)} \text{ s, working in metres}$$
$$= 2\pi \times 0.16$$
$$= 1\text{s}$$

Frequency = 1/*t* = 1/1 = 1 Hz, or 60 vibrations per minute.

Motor car suspension systems are designed to have the following approximate values for frequencies:

- Front suspension – 60/min to 80/min
- Rear suspension – 70/min to 90/min.

6.3 Valve spring surge

When the natural frequency of vibration of a spring coincides with the frequency of a disturbing force, a condition known as resonance occurs, and the amplitude of vibration becomes large and uncontrollable. This occurs when an engine speed exceeds a safe speed and valve surge (bounce) takes place. For example, in a 4 stroke engine at a speed of 6000 rev/min, the valves are opened and closed 3000 times a minute, which is a frequency of 50 cycles per second. In practice the valve springs are designed to have a frequency much higher, consistent with their being not so strong that they stress the camshaft and valve operating mechanism.

Valve springs

The variable pitch valve spring on the left of Figure 6.3 is designed so that the coils are compressed against each other progressively, with the coils closest to the cylinder head being compressed first. This means that the number of active coils varies with the opening and closing of the valve. The effective stiffness and frequency of vibration thus varies during valve operation which helps to minimise the risk of valve surge occurring during normal engine operation.

The double springs shown on the right of Figure 6.3 are of different stiffness, and it is claimed that the resonance of one spring is cancelled out by the other spring, which has a different resonant frequency. Another advantage of the double spring is that, in the event of one spring failing, the other may allow the valve to continue to function.

Example 6.3

A poppet valve is designed to have simple harmonic motion. The full lift of the valve is 7 mm and the time taken to open and close the valve, at a certain engine speed of 0.01 second. The valve has a mass of 0.15 kg. Calculate the inertia force at each end of the valve's motion.

Close coil at
cylinder head end

Two springs one
inside the other

Figure 6.3 Anti-surge valve springs

Solution 6.3

Periodic time $t = 0.01 = \dfrac{2\pi}{\omega} \therefore \omega = \dfrac{2\pi}{0.01} = 200\pi \dfrac{\text{rad}}{\text{s}}$.

Angular acceleration $\alpha = \omega^2 x$ where $x =$ half the lift

$$\therefore \alpha = (200\pi)^2 \times 0.0035 = 1382 \dfrac{\text{rad}}{\text{s}}$$

Force on valve $= m\alpha = 0.15 \times 1382 = 207.3$ N

6.4   Analysis of piston motion

The following analysis of piston motion highlights a number of features of engine construction and operation that lead to disturbing forces that cause vibrations

Figure 6.4 shows the layout of a simple reciprocating engine mechanism.

Let $r =$ length of the crank throw,
$\omega =$ angular velocity of the crank,
$l = n.r =$ length of connecting rod ($n =$ ratio conn rod length/crank throw, $n = \dfrac{l}{r}$),
$\theta =$ crank angle in relation to top dead centre,
$x =$ piston displacement from top dead centre position,
$v_p =$ velocity of the piston,
$a_p =$ acceleration of the piston.

With reference to Figure 6.4, the line CR is perpendicular to the line of stroke.

The displacement of the piston

$$x = P_1 P = P_1 O - PO = l + r - (l \cos \varphi + r \cos \theta)$$
$$= r(1 - \cos \theta) + l(1 - \cos \varphi) \tag{1}$$

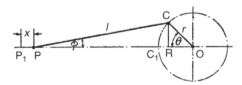

Figure 6.4 Piston motion

But

$$CR = r \sin \theta = l \sin \varphi$$

$$\therefore \sin \varphi = \left(\frac{r}{l}\right) \sin \theta = (\sin \theta)/n \tag{2}$$

$$\cos \varphi = \sqrt{(1 - \sin^2 \varphi)} = \sqrt{\left(1 - \frac{\sin^2 \theta}{n^2}\right)}$$

$$= \left(\frac{1}{n}\right) \sqrt{(n^2 - \sin^2 \theta)} \tag{3}$$

Substituting in equation (1) produces the following;

$$x = r(1 - \cos \theta) + nr \left\{ 1 - \left(\frac{1}{n}\right) \sqrt{(n^2 - \sin^2 \theta)} - \right\}$$

$$x = r \left\{ 1 + n - \cos \theta - \sqrt{(n^2 - \sin^2 \theta)} - \right\} \tag{4}$$

Differentiating with respect to time to give piston velocity,

$$v_p = \frac{dx}{dt} = \frac{d\theta}{dt} \cdot \frac{dx}{d\theta} = \omega r \left\{ \sin \theta + \frac{\sin 2\theta}{2\sqrt{(n^2 - \sin^2 \theta)}} - \right\} \tag{5}$$

For piston acceleration a_p, the velocity is differentiated with respect to t:

$$a_p = \frac{dv_p}{dt} = \frac{d\theta}{dt} \cdot \frac{dv_p}{d\theta} = \omega^2 r \left\{ \cos \theta + \frac{n^2 \sin^2 \theta + \sin^4 \theta}{(n^2 - \sin^2 \theta)^{\frac{3}{2}}} - \right\} \tag{6}$$

Because $\sin^2 \theta$ is small compared with n^2, the equations for piston velocity and acceleration may be written as follows to give accurate approximate values:

$$v_p \approx \omega r \left\{ \sin \theta + (\sin 2\theta)/2n \right\} \tag{7}$$

$$a_p \approx \omega^2 r \left\{ \cos \theta + \frac{\cos 2\theta}{n} \right\} \tag{8}$$

Forces on the piston and part of the connecting rod.

The inertia forces on the piston are obtained by multiplying the piston mass by the acceleration; $F_p = \text{mass} \times \omega^2 r \left\{ \cos 2\theta + \dfrac{\cos 2\theta}{n} \right\}$.

$$(9)$$

6.5 Primary and secondary forces

For engine balance purposes, it is useful to consider equation 6.9 as two separate parts. One part is called the primary force: $F_p = m\omega^2 r \cos\theta$. The other is called the secondary force: $F_s = m\omega^2 r \left\{ \dfrac{\cos 2\theta}{n} \right\}$.

These two expressions show that the maximum value of the secondary force is only 1/n of the maximum value of the primary force, but that this maximum value occurs four times per revolution of the crankshaft, as compared with twice per revolution of the crankshaft for the maximum primary force. A graphical presentation of primary and secondary forces is shown in Figure 6.5.

Example 6.4

A reciprocating engine has a stroke of 100 mm and a connecting rod length of 220 mm. Determine the piston velocity and acceleration at an engine speed of 3000 rev/min when the crank is at an angle of 30° past top dead centre.

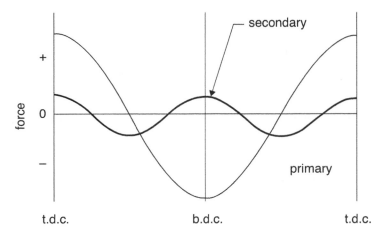

Primary and secondary forces on one stroke of piston

Figure 6.5 Comparison of primary and secondary forces

Solution 6.4

$$n = \frac{l}{r} = \frac{220}{50} = 4.4$$

velocity of piston $v_p = \omega r\left\{\sin\theta + \left(\frac{\sin 2\theta}{2n}\right)\right\} m/s$

$$\omega = \frac{3000}{60} \times 2\pi = 314.2 \, rad/s$$

$$v_p = 314.2 \times 0.05\left(0.5 + \frac{0.866}{2 \times 4.4}\right) = 9.4 \text{ m/s}$$

Acceleration of piston $\alpha_p = \omega^2 r\left(\cos\theta + \frac{\cos 2\theta}{n}\right)$

$$\therefore \alpha_p = 314.2^2 \times 0.05\left(0.8660 + \frac{0.5}{4.4}\right) = 98721.6 \times 0.05 \times 0.98 = \frac{4837m}{s^2}.$$

Example 6.5

If the mass of the piston assembly in Example 6.4 is 0.8 kg, calculate the primary and secondary inertia forces acting on it.

Solution 6.5

Primary force $= \text{mass} \times \omega^2 r \cos\theta = 0.8 \times 98721.6 \times 0.05 \times 0.866$

$$= 3.42 \text{ kN}$$

Secondary force $= F_s = m\omega^2 r\left\{\frac{\cos 2\theta}{n}\right\}.$

Secondary force $= 0.8 \times 98721 \times 0.05 \times \frac{0.5}{4.4} = 448.73 \text{ N}$

6.6   Balance of primary and secondary forces in a 4 cylinder in-line engine

In the 4 cylinder in-line engine the reciprocating parts are identical for each cylinder and the cranks are arranged to provide uniform firing intervals. In most cases the distances between the cylinders are equal. In this design the primary forces on the rising and falling pistons are equal, which means that primary force balance is achieved within the engine. In some engines secondary and harmonic forces are not balanced, and this calls for the use of a secondary force balancer (shown in Figure 6.6) which makes use of two shafts with balance weights attached that rotate at *twice* the crankshaft speed.

By reference to the graph in Figure 6.5, it can be seen that the secondary force is at a maximum four times per revolution of the crank which is when $\cos 2\theta = 1$, *or* -1.

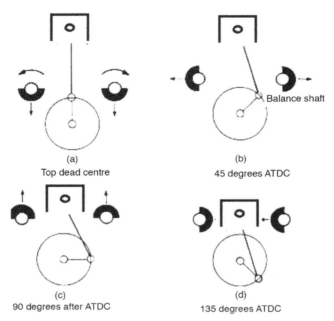

(a)
Top dead centre

(b)
45 degrees ATDC

Balance shaft

(c)
90 degrees after ATDC

(d)
135 degrees ATDC

Twin shaft secondary force balancer

Figure 6.6 Secondary force balancer

6.7   Force diagrams – engine balance

The mathematical analysis of piston motion shows that the disturbing force due to a reciprocating mass is identical with the component parallel to the line of stroke of the centrifugal force produced by an equal mass attached to, and revolving with, the crankpin. Problems on the primary balance of reciprocating masses may therefore be solved by the use of vector diagrams. In order to prove complete balance of an in-line engine, the force and couple diagrams must be closed figures.

Example 6.6

A 6 cylinder, 4 stroke, in-line engine, firing order 1, 4, 2, 6, 3, 5, has a stroke of 75 mm and a connecting rod length between centres of 150 mm. The distance between the cylinders is equal at 100 mm. The piston assembly has a mass of 400 grams.

(a)   Calculate the maximum inertia force on each piston at an engine speed of 3000 rev/min.

(b) Draw vector diagrams to show the state of balance of primary and secondary forces and couples. Use the point midway between cylinders 2 and 3 as reference plane.

Solution 6.6

All of the masses are rotating at the same speed which means that all of the force vectors are of the same length. The primary and secondary force polygons are closed figures showing complete balance of these forces. As far as couples are concerned, the right half of the crankshaft is a mirror image of the left half, and the couples (moments) to the right are equal and opposite to those on the left, which indicates that couples are also balanced. See Figure 6.6(b).

(a) Maximum inertia force occurs at the end of the stroke where $\cos \theta$ and $\cos 2\theta = 1$.

$$\text{Inertia force} = mr\omega^2 \left(\cos \theta + \frac{\cos 2\theta}{n} \right)$$

$$\omega = 2\pi \times \frac{3000}{60} = 314.2 \ radians/sec$$

$$n = \frac{l}{r} = \frac{150}{37.5} = 4$$

$$\therefore \text{inertia force} = 0.4 \times 0.0375 \times 314.2^2 \left(1 + \frac{1}{4} \right) = 1.85 \text{ kN}$$

(b)

Cylinders 1 to 6 in-line

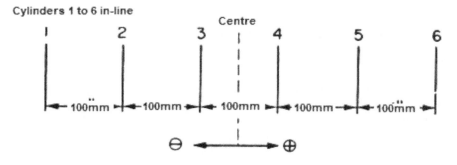

Figure 6.6(b) Force diagrams

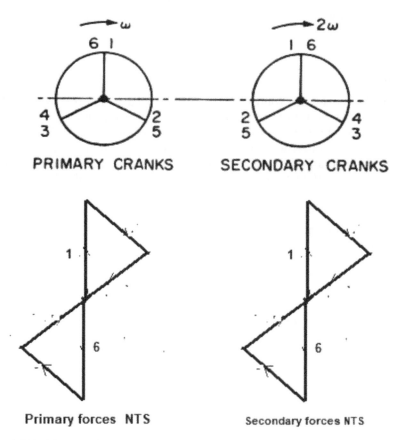

Figure 6.6(b) (Continued)

6.8   Torsional vibration of crankshaft

Figure 6.7 shows how the crankshaft torque varies while the engine is rotating. The space between the peaks on the graph represents a period of time in which the flywheel is accelerated and slowed down. During this period a small angle of twist is induced in the crankshaft at a frequency (number of times per second) that corresponds to the number of firing impulses each second. When the frequency of vibration matches the natural frequency of vibration of the crankshaft, resonance occurs and the degree of twist of the crankshaft increases. Figure 6.8 shows the effect that resonance has on engine noise level. The mechanical force generated

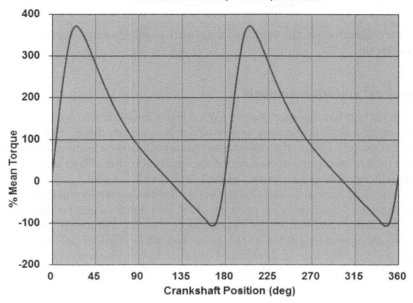

Figure 6.7 Torque fluctuation between power strokes

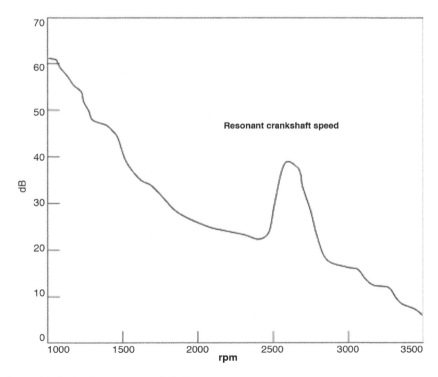

Figure 6.8 Peak vibration in crankshaft

by the vibration at this engine speed can cause damage in the vehicle powertrain.

Torsional vibration dampers

In the viscous film damper (Figure 6.9), relative movement between the inertia ring and the outer casing is resisted by the silicone fluid. The shearing action that takes place absorbs energy that would otherwise twist the crankshaft thus reducing the occurrence of torsional vibration. These dampers are normally fitted to the front end of the crankshaft.

In the dual mass flywheel (Figure 6.10), the secondary flywheel mass can rotate slightly relative to the primary mass, and it does so under the control of the springs and the flange. This action absorbs disturbing forces that would otherwise be passed to the gears and other parts of the drive train. This is of significance in powerful 4 cylinder in-line engines where the disturbing forces can be quite large and lead to powertrain damage and unacceptable noise.

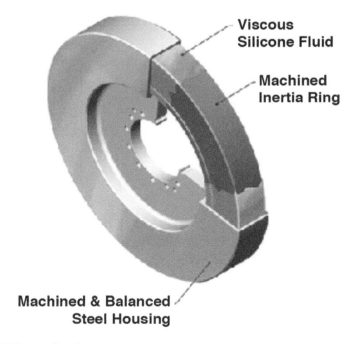

Viscous
Silicone Fluid

Machined
Inertia Ring

Machined & Balanced
Steel Housing

Figure 6.9 Viscous film damper

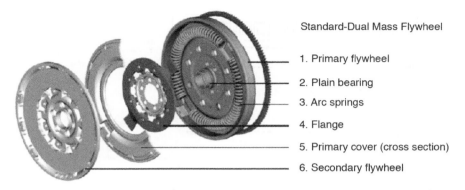

Standard-Dual Mass Flywheel

1. Primary flywheel

2. Plain bearing

3. Arc springs

4. Flange

5. Primary cover (cross section)

6. Secondary flywheel

Figure 6.10 Dual mass flywheel

6.9   Critical (whirling) speed of drive shaft

Whirling speed

This is with reference to Figure 6.11 that can represent a propeller shaft. The distributed load causes the beam (shaft) to sag (deflect), as shown in an exaggerated form in Figure 6.11(b). As the shaft rotates in its deflected form, the *centrifugal force* that arises is because the centre of mass is now at a small radius from the original axis. The centrifugal force is resisted by a force that is produced by the *elastic properties* of the shaft material; these forces are equal and opposite in direction. Both the centrifugal force and the elastic force are proportional to the deflection. As the speed of shaft rotation is increased the magnitude of the elastic force becomes indeterminate, the shaft becomes unstable and it is said to whirl. The speed at which this occurs is called the *whirling speed*, or the *critical speed*, and it is equal to the natural frequency of vibration of the shaft, or beam. For a simply supported beam, the whirling (critical) speed is given by the formula $f = \dfrac{1.57}{l^2}\sqrt{\dfrac{EI}{m}}$, where $l = $ length of shaft, E is the modulus of elasticity, I is the second moment of area, m is the mass length and l is the length. This expression shows that the whirling speed is inversely proportional to shaft length – the longer the shaft, the lower the whirling speed. In most cases the question of whirling and critical speed arises in connection with propeller shafts of front engine rear drive vehicles. The divided propeller shaft, as shown in Figure 6.12 is a method of construction that is used to overcome any likelihood of whirling and consequent damage to the drive train.

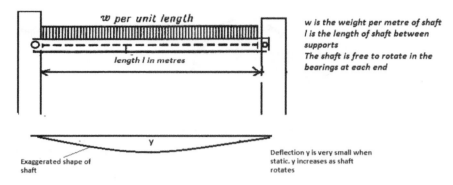

Figure 6.11 Simple beam carrying uniform load along its length

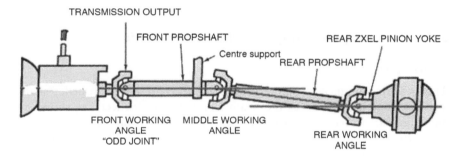

Figure 6.12 A divided propeller shaft.

Example 6.7

A simply supported beam (shaft) 2 m long has a moment of inertia $1.5 \times 10^8 \, \text{mm}^4$, and it carries a uniformly distributed load of 1500 kg/m. Find the natural frequency, taking $E = 207 \, \text{kN} / \text{mm}^2$.

Solution 6.7

$$\text{Frequency } f = \frac{1.57}{4} \sqrt{\frac{207000 \times 1.5 \times 10^8}{1500 \times 10^6}} = 20.4 \text{ per sec}$$

Practical method for obtaining critical speed

The nomogram in Figure 6.13 is based on theory and empirical evidence and it provides details of critical speeds of types of shaft produced by Rockwell.

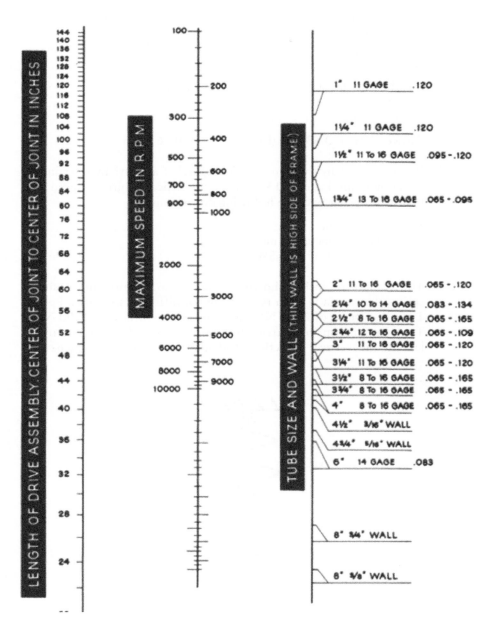

Figure 6.13 Nomogram (chart) for critical speed (Rockwell)

Self-assessment questions

6.1   A single cylinder engine has a bore of 90 mm and a stroke of 100 mm. The gas pressure on the power stroke is 60 bar at a crank angle of 40 degrees past TDC. The connecting rod length is 220 mm between centres. Determine:

(a) the side thrust of the piston on the cylinder wall
(b) the force in the connecting rod
(c) the torque on the crank.

Answer 6.1: (a) 3.63 kN, (b) 25.7 kN, (c) 0.96 kNm

6.2   What is meant by the term *whirling speed* of a shaft? By referring to the nomogram, Figure 6.13, estimate the critical (whirling) speed of a hollow propeller with the following dimensions:

• Length 1.5 m
• Outside diameter 75 mm
• Thickness of tube 16 SWG.

6.3   The composite spring shown in Figure S.6.3 represents a single spring that is made from two springs of different strengths. Spring 1 has a stiffness of 2200 N/m and spring 2 a stiffness of 1000 N/m. The combined springs carry a weight of 80 N and it is free to oscillate in the vertical direction. Calculate the natural frequency of vibration.

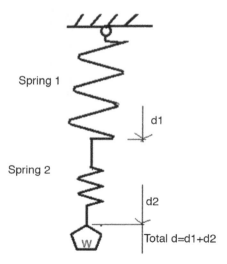

Figure Answer 6.3 Frequency = m 1.46 Hz.

6.4   A reciprocating engine has a stroke of 100 mm and a connecting rod length of 240 mm. Determine the piston velocity and acceleration at an engine speed of 3000 rev/min when the crank is at an angle of 30° past top dead centre.

Answer 6.4: piston velocity 9.3m/s, acceleration 4744 m/s²

6.5   A petrol engine has a bore diameter of 98 mm and a stroke length of 110 mm. The connecting rod is 200 mm between centres and the pistons have a mass of 0.750 kg. Calculate the inertia force on the piston at the TDC position when the engine speed is 4800 rev/min.

Answer 6.5: inertia force on piston = 10.42 kN

6.6   An engine has poppet valves that operate with simple harmonic motion. The lift of the valve is 5 mm, and at a certain engine speed it takes 0.006 seconds to open and close the valve. The valve weighs 0.140 kg. Calculate the force acting on the valve at each end of its stroke.

Answer 6.6: force on valve = 383.6 N

6.7   A 3 kg weight is suspended from a vertically mounted spring and it stretches the spring by 14 mm.

(a)   Calculate the number of complete oscillations per second
(b)   If the weight is displaced a further 25 mm below the rest position and then released, calculate the maximum force in the spring.

Answer 6.7: (a) 4.21 Hz, (b) 81.3 N

Chapter 7

The CAN system

7.1 Networked systems on vehicles

Computer controlled vehicle systems such as engine management, traction control, anti-lock braking and stability control need to communicate with each other in order to obtain the optimum performance from the vehicle. To achieve this end the individual systems are linked together (networked) by a communication bus that permits data to be interchanged between the systems at a very high data transfer speed. This is displayed in Figure 7.1. Systems that operate at high speeds of data transfer are said to operate in **real time**.

Bus

Bus is the name given to the wires or fibre optics that transmits the data used to exchange information between the computer systems that operate together to control vehicle behaviour. A typical example of a vehicle system that uses a data bus is traction control where the engine management system

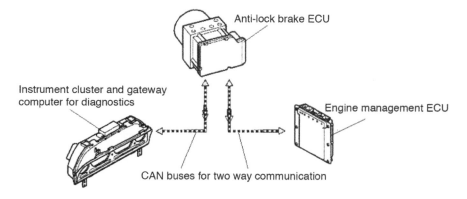

Figure 7.1 CAN on traction control

and the anti-lock braking system act together to prevent wheel spin. A data bus on a vehicle may be a single wire, a fibre optic thread or, in the case of CAN, a pair of wires.

Serial data transmission (Baud rate)

In a simple data transmission system, the data (computer words) consists of a series of noughts (0s) and ones (1s) that represent voltages. These 0s and 1s are called bits, and they are transmitted serially (one after the other) along the wires that connect parts of the system together. The rate at which the 0s and 1s are transmitted is called the Baud rate; for example 10 bits per second is 10 Baud.

CAN – controller area network (Robert Bosch)

The CAN system of data transfer for networking of electronic control units was developed by Robert Bosch Gmbh in the 1980s and it has now become widely used in automotive practice. The basic CAN data bus consists of two wires which transmit the data signals in the form of voltages.

Two types of CAN are commonly used on vehicles:

1   **Low speed CAN** which is used for data transfer speed of up to 125 kbaud. This is suitable for the control of body systems such as seat adjustment, air conditioning and other systems related to driver and passenger comfort.
2   **High speed CAN** which is used for data transfer speeds of up to 1 Mbaud. High speed CAN is used for systems such as engine management, traction control and transmission control.

7.2   Data transmission rates on CAN buses

Data transmission rates are measured in bits/s or baud; 1 baud = 1 bit/s. The maximum bit rate is dependent on the length of the CAN bus. The data transmission figures shown in Table 7.1 are those that are specified by ISO 11898.

Table 7.1 Data transmission

Maximum permitted bit rate	Max length of CAN bus
1 Mbits/s	40 m
500 kbits/s	100 m
250 kbits/s	250 m
125 kbits/s	500 m
40 kbits/s	1000 m

7.3   CAN voltages

The logic levels, namely binary 0 and 1, that are used for computing purposes are described as dominant and recessive respectively.

The **dominant** state represents a binary 0, and the **recessive** state a binary 1. The voltages that constitute the binary 0 and 1, outlined in Table 7.2, are made up by the difference between the voltage on the H line and that on the L line.

Table 7.2 CAN voltages

High speed CAN-C	CAN_H Line	CAN_L Line	Logic level	Voltage
Dominant state	3.5 volt	1.5 volt	Binary 0	2 volt
Recessive state	2.5 volt	2.5 volt	Binary 1	0 volt

Low speed CAN_B	CAN_H Line	CAN_L Line	Logic level	Voltage
Dominant state	3.6 volt	1.4 volt	Binary 0	2.2 volt
Recessive state	0 volt	5 volt	Binary 1	5 volt

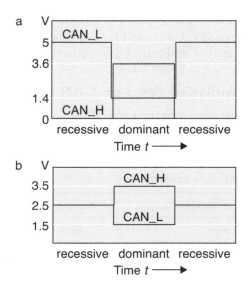

Figure 7.2 CAN_H and CAN_L

7.4 The CAN bus

The two wires that are used to make up the CAN bus are twisted together (Figure 7.3) throughout their length and are normally incorporated into the wiring loom. This method of forming the bus is known as a twisted pair, and it is used to minimise the risk of electrical interference that can occur when data is being transmitted.

In systems where the CAN bus operates at speeds in excess of 500 kb/s, it is considered necessary to use screened cables to reduce the risk of interference with other systems.

Bus layout

For study purposes the CAN bus is represented by a pair of parallel lines as shown in Figure 7.4.

The ECUs that are networked are called nodes. Each node is connected to the two bus wires that are known as CAN_H and CAN_L.

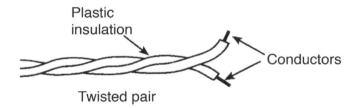

Figure 7.3 A twisted pair

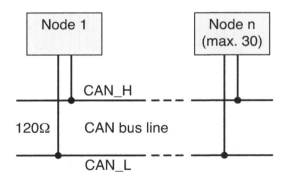

Figure 7.4 CAN bus nodes

7.5   The CAN node

The CAN nodes contain a microcontroller which is a computer that is able to perform a range of functions. Each node is able to transmit and receive data, and this is achieved by means of a device called a transceiver. Each message on the bus is read by all nodes, but the message is used only by those nodes that require it. This is achieved by means of an identifier code that is incorporated into the message.

Should two nodes attempt to connect to the bus simultaneously, there is a risk that a collision may occur resulting in incorrect working of systems. In order to prevent this from happening, a system known as "wired AND" is used. In effect, when the bus has a binary 0, it will accept only an input of the same value. In general, if two messages are trying to access the bus simultaneously, the one whose identifier has the lowest binary value will be given access.

7.6   Access to the CAN bus

The identifier C in Figure 7.5 is a binary number that represents the unit that is addressing the bus. When two units are trying to access the bus simultaneously, the identifier with the lowest binary value will be given priority. For example, 00000001 will be given priority over 00010001.

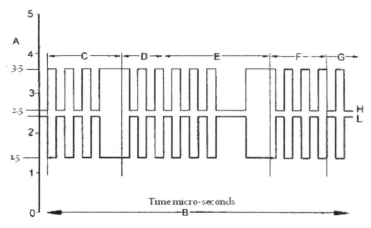

A. = Voltage; **B.** = Time; **C.** = Identifier; **D.** = Length; **E.** = Message; **F.** = Error check; **G.** = Next message; **H.** = CAN High; **L.** = CAN Low

Figure 7.5 A pattern of the voltage levels in a CAN message.

Error protection

Because vehicle systems operate under a wide range of hostile conditions such as varying, temperatures – very hot and very cold – and varying voltages, there is a risk of errors. To guard against this, checks are made, and when an error in data occurs, CAN devices keep a count of the number of times that the error occurs. CAN devices are able to distinguish between temporary errors and permanent ones. When a device is deemed to be permanently defective, it disconnects itself electrically from the network. Under such conditions the remainder of the network continues to function.

Cyclic redundancy check (CRC): error detection

Messages that are transmitted to the data bus are checked by the transmitting device and the receiving device using a mathematical process that incorporates a cyclic redundancy code. If the message at the receiving end does not match that sent by the transmitting device, the message is rejected.

7.7 Types of CAN

High-speed CAN, CAN-C, conforms to ISO Standard 11898-2 and operates at 125 kbits/s up to 1 Mbits/s. It is used for those systems that operate in real time, such as:

- Engine management
- Transmission control, including traction control
- Electronic and anti-lock braking systems
- Vehicle stability systems
- Instrument cluster.

Low-speed CAN, CAN-B, is covered by ISO Standard 11898-3 and operates at data transmission speeds of 5 kbits/s up to 125 kbits/s. It is used for comfort and convenience systems and others in the body area such as:

- Air conditioning
- Lighting system
- Seat adjustment
- Mirrors
- Sliding sun-roof
- Doors.

7.8 Number of nodes on a CAN network

- Should one system that is connected to the CAN bus fail, the other systems on the bus continue to function.
- Systems can be added.

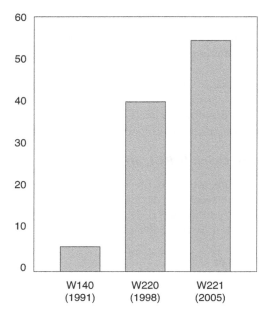

Figure 7.6 Approximate number of networked ECUs (Bosch)

Figure 7.6 shows how the number of systems that are networked on a vehicle is likely to continue growing.

Multiplexing

The term multiplexing is used in connection with several different electronics applications. Some of these are;

7.9   Multiplexed messages on a data bus

Time division multiplexing is a system of data transmission where digital messages are transmitted on a bus (wire) in time slots, one message following another. The process happens at high speed and gives the impression that the messages are being transmitted simultaneously.

Multiplexed wiring

The main aim of multiplexed wiring systems is to reduce the amount of wire that is used on a vehicle. Instead of having wires running from each switch on the control panel to each individual item, a data bus is used to transmit messages around the vehicle. All commands such as "switch on the headlights" are transmitted on a data bus and are acted upon by the headlight system alone. The same bus is used for other systems like screen wipers and seat and mirror adjustment.

7.10 Data transmission

Sampling rate

Analogue signals from sensors are read by the control module at frequent intervals. The rate at which they are read is called the sampling rate, and the greater the accuracy required the faster will be the sampling rate. In some engine management systems the signal from the engine speed sensor is read on each firing stroke and the frequency of readings is dependent on the number of cylinders in the engine.

- In a 4 cylinder, 4 stroke engine running at 6000 rev/min, the number of firing strokes per second $= \frac{4}{2} \times \frac{6000}{60} = 200/s$. This is equal to an interval of 5 m/s.

The amount of activity on a data bus can be quite large and the speed of data transmission is a factor that is considered in their design. Electrical magnetic signals such as light, infrared and radio forms of communication travel at the speed of light which is 3×10^8 m/s in a perfect vacuum and slightly less in air.

Various forms of data transmission are used in motor vehicles. Typical examples being:

- Electrical pulses for serial data transmission on copper cables
- Light pulses on fibre optics
- Wireless communication such as satellite navigation, remote controls and collision avoidance.

7.11 Losses and interference

Copper is a good conductor of electricity, and it is the material that is often used for the buses on vehicle systems. Unfortunately, copper is not a perfect conductor, and some electrical energy can be lost in transmission. The effect of the loss of signal strength is that the signals may be impaired in some way. The main problems associated with data signals on copper conductors are:

- **Crosstalk.** This happens when electrical signals in other conductors nearby interfere with the signals on the bus. The twisted pair is one method of minimising this effect.

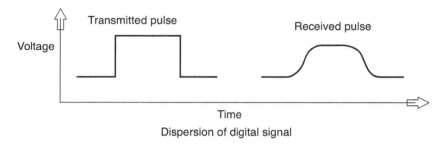

Figure 7.7 Digital signal dispersion.

- **Attenuation.** The signal becomes weaker as it travels along the conductor. It loses a constant fraction of its size along a given distance. If the signal is 90% of its original size at 10 m along the conductor, it will be 81% (90% × 90%) of its original size at 20 m and so on.
- **Dispersion.** Data signals can change shape as well as size during transmission. A pulse that starts transmission as distinct square wave form with sharp edges can lose its sharp edges by the time that it reaches the receiving end. The effect of dispersion is shown in Figure 7.7 where the clear definition of the digital signal is slightly distorted

7.12   Other data buses on vehicles

The K bus is a two way bus that is used for diagnostic purposes, and it is the bus through which fault codes can be read and signals can be sent that allow some actuators to be tested. The K bus operates at Baud rates up to 10.4 kb/s.

Some proprietary systems such as the Land Rover DS2 operate at 9.6 kb/s.

7.13   Diagnostics

The gateway

The gateway is a computer that can access most data on the various data buses, as shown in Figure 7.8. Its main function is to translate (decode) the various signals so that they can be read by diagnostic equipment such as a scan tool.

Access to diagnostic data such as fault codes is achieved through the standardised diagnostic connecter that is placed in an accessible position in the

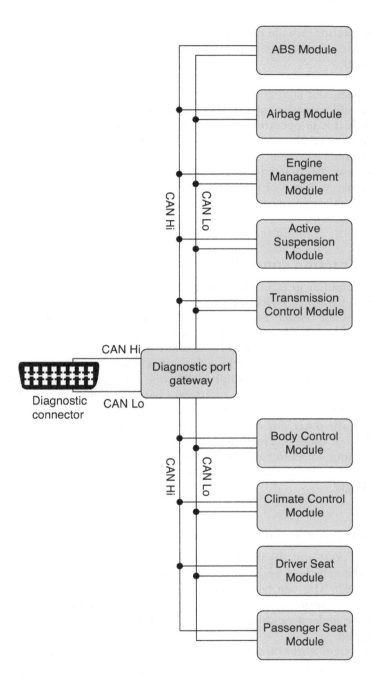

Figure 7.8 The gateway (Omitec Ltd)

vehicle interior. A wide range of diagnostic test equipment is available, and this enables a technician to perform a number of functions that includes the following:

- Reading out fault codes.
- Examining aspects of a system's behaviour such as engine management, while it is in operation. For example, individual actuators such as fuel injectors can be activated to see if their performance meets previously determined criteria.
- In cases restricted to manufactures agents (main dealers), a process known as "application engineering" can be performed. This process allows the approved agent to alter performance features that were set at the factory.

7.14   Traction control

The differential gear in the driving (live) axle of a vehicle permits the wheel on the inside of a corner to turn more slowly than the wheel on the outside of the corner. For example, when a vehicle is turning sharply to the right, the right-hand wheel of the driving axle will rotate very slowly and the wheel on the left-hand side will rotate faster. However, this same differential action can lead to a loss of traction and wheel spin under other circumstances. If for some reason one driving wheel is on a slippery or loose-gravel surface when an attempt is made to drive away, the wheel on the slippery surface may spin while, because of differential action, the wheel on the other side of the axle will stand still. This can lead to a complete loss of motion. Traction control enables the brake to be applied on the spinning wheel to stop the spinning and simultaneously, via the ECU, operates the electronically controlled throttle to regulate engine power accordingly.

ESP (electronic stability programme)

Figures 7.9(a) and (b) illustrate the effect of stability control. In Figure 7.9(a) the vehicle is veering from the steered path in a condition known as oversteer. This can be corrected by application of the inside rear brake and possibly some easing of the throttle by the engine ECU. Figure 7.9(b) shows the understeer condition; here the brakes on the outside of the turn are applied.

This system works on vehicles equipped with electronic brake distribution where individual wheel brakes can be applied automatically. Such vehicles may also be equipped with a yaw sensor which detect the angle yaw (twist) of the vehicle about the vehicle's vertical axis.

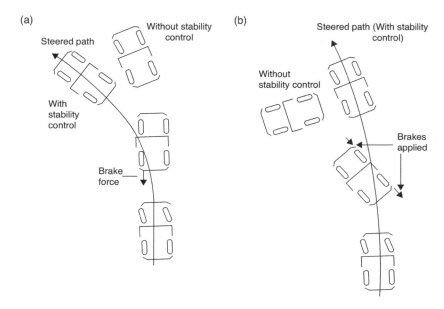

Figure 7.9(a, b) Stability control

7.15   Adaptive cruise control

Adaptive cruise control uses radar and, in some cases, a camera to ascertain the distance between a moving vehicle and a vehicle in front (moving in the same direction). After the driver selects the desired cruising speed, the radar signals, as recorded by the ECU, are used to determine what action if any is required to maintain speed and a safe distance between vehicles. If *acceleration* is required, the cruise control ECU connects with the engine ECU (via CAN) to increase fuelling in a diesel engine, or open the throttle in a petrol engine. If the speed of the vehicle exceeds the set speed or the vehicle is getting too close to the one in front the system will, through the engine ECU, intervene to slow the vehicle down. If the situation is severe, such as the vehicle in front making an emergency stop, the brakes may be applied through the brake system ECU. This action is facilitated because wheel brakes can be operated via CAN without driver action. A later development that is used in some trucks makes use of GPS. The cruise control settings in the map in the control unit are designed to take advantage of GPS information so that the system can take account of the terrain in different situations. In the event of an emergency such as a vehicle cutting in front of another, the required level of braking may be higher than that provided by the adaptive system, in which case the driver is alerted, by audible signal, to apply the brakes accordingly.

7.16   Hill stop and start

The hill stop and start system relies on CAN to coordinate actions between the engine control and the braking system. With this system the engine is shut down as soon as the vehicle is stationary, and it restarts by depressing the clutch or, in the case of automatic transmission, when the foot brake is released. The use of this system provides an improvement in fuel consumption and a reduction in harmful emissions.

Self-assessment questions

7.1   The following are the binary codes of the identifiers of 4 different nodes that are attempting to access the CAN bus simultaneously. Which one will be given priority?

    (a)   00001111
    (b)   00011001
    (c)   00110000
    (d)   00010101

7.2   Above what Baud rate is it considered necessary to use screened cables?

7.3   At what speed do electrical pulses travel along a copper cable?

7.4   What was the approximate number of networked ECUs on a Mercedes car in 2005?

7.5   An engine management controller reads the engine speed sensor on each firing stroke. Calculate the time in milliseconds between each firing stroke on a 6 cylinder, 4 stroke engine when the engine speed is 3000 rev/min?

    Answer 7.5: 150 per second

7.6   What Baud rate is used on the diagnostic bus?

7.7   What is meant by the term *attenuation*?

7.8   Which systems are likely to operate at CAN data speed of 125 kb/s up to 1 Mb/s?

7.9   What are the voltage levels on the CAN_H line?

7.10   What is the value of the recessive voltage on CAN_H?

7.11   What is the maximum length of cable that can be used in 1 Mb/s CAN system?

Chapter 8

Engine mapping

An overview

The data stored in the individual cells of the ECUs memory can be represented graphically in the form of a map. Information for constructing the map is obtained by conducting a series of tests on the engine, and the program for these tests is called engine mapping. These tests determine the performance and investigate the effects of each variable that has some bearing on the output of the engine. When the effects are known, the settings that give the best performance can be determined and recorded in ECU memory in various forms, including lookup tables.

A microcontroller is the central component of an ECU. In order to perform computations, the microcontroller has a program which is stored in the read only memory (ROM). The ROM memory contents are defined during manufacture, and they cannot be altered. The ROM has limited capacity and is often supported by additional ROM in the case of complicated applications.

The RAM (read/write) working memory is needed to hold variable data such as sensor inputs. The RAM is volatile, which means that all data is lost when the electricity supply is removed. However the RAM is connected in such a way that it remains in use when the ignition is switched off. This means that when the engine is restarted, the control unit has access to adaptation data (learned data relating to engine condition). If the battery is disconnected, the data in RAM will be lost.

The EEPROM is an electrically erasable read only memory that holds, among other data, the code for the immobiliser and important adaptation data which must remain electrically powered when the ignition is switched off.

The fact that it is electrically erasable means that every single memory location and its contents can be erased and replaced. So any data in the EEPROM is amenable to reprogramming with suitable equipment and information. The software calibration process, shown in Figure 8.1, is as follows:

- **Defining the desired characteristics.** This is done by the ECU manufacturer in consultation with the engine designer.

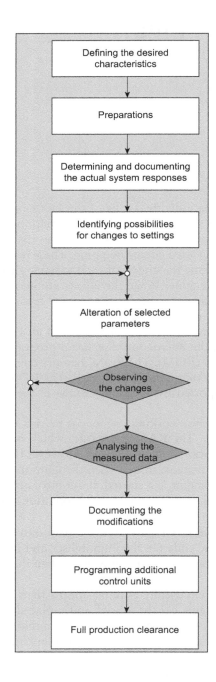

Figure 8.1 Mapping

- **Preparations.** Special ECUs are used for calibration purposes. Compared with normal production ECUs, some of the parameters can be altered.
- **Determining and documenting the actual system responses.** The screen shown in Figure 8.2 shows how the information about various settings can be displayed in the form of graphs (maps) and tables. The measurement data can be viewed while work is being carried out. This permits the engine response to changes to be investigated while they are taking place.
- **Identifying possibilities for changes to settings.** With the help of the ECU software it is possible to identify which parameters are best suited to altering system response in the desired manner.
- **Alteration of selected parameters.** All parameters can be altered while the engine is running so that the effects of the changes can be measured and observed on the computer screen.
- **Analysing the measured data.** This stage involves comparing and documenting the changes and their effects. This stage is important because several people will be involved in the process of engine optimisation at different times.
- **Documenting the modifications and programming additional control units.** This applies to large scale manufacturing.

The computer display shows a range of data that is displayed while dynamometer tests are in progress. In this display there are some line graphs, a look-up table and a three-dimensional graph. The ECUs that are used for initial mapping have settings that can be altered, and the effect of any

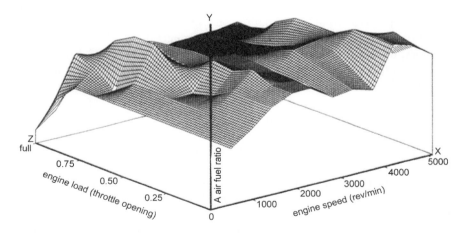

Figure 8.2 Petrol engine fuelling map

changes that are made can be displayed instantly. This procedure relates to the following stages in Figure 8.1.

- Identifying possibilities for changes to settings
- Alteration of selected parameters.

8.1   Stages of the engine mapping process

The mapping process takes place in two stages:

1   On a dynamometer in a test cell. The test cell is designed so that the results of tests can be shown to conform to various international standards about performance, exhaust emissions, etc.
2   On a specially designed test track. At this stage the map settings are put through a range of tests that simulate road going conditions

Interpreting map data
Fuelling

Figure 8.2 shows a petrol engine fuelling map, which is sometimes called the lambda (λ) map. The variables shown on this map are:

- Air-to-fuel ratio λ on the Y axis
- Engine load as indicated by throttle valve position on the Z axis
- Engine speed in rev/min on the X axis.

The data from which the map is produced is acquired by extensive testing on a fully instrumented dynamometer test bed. Eventually the data is converted to computer code and then stored in the ECU memory. Examination of the map reveals that many data points are stored across the operating range of the engine. Each point on the map is encoded and stored in ECU memory, in form of a lookup table such as the one in Figure 8.3. For each amount of throttle opening and corresponding engine speed, there is a λ setting that determines the amount of fuel required.

Effect of air-fuel ratio on operating variables

Figure 8.4 shows how torque and combustion products vary as the air-to-fuel ratio (λ) is changed. The term stoichiometry is used to refer to the air-to-fuel ratio when combustion is chemically correct. At stoichiometry the air-to-fuel ratio is approximately 14.7:1, and this is taken as $\lambda = 1$; for rich mixture, $\lambda < 1$, and for weak mixtures, $\lambda > 1$.

When considering what if any changes to map settings are to be made, it is necessary to understand their effect on other variables such as fuel

1995 5.0 Mustang Base OL Fuel Table (A/F Ratio)

LOAD	500	700	900	1100	1300	1500	2000	2500	3000	4000
0.90	8.46	8.01	9.49	10.29	12.35	12.35	12.35	12.01	12.01	12.01
0.80	8.46	8.46	9.95	10.52	12.92	13.15	13.50	13.61	13.61	12.92
0.70	8.81	8.81	10.41	10.98	13.73	13.95	13.95	14.30	14.30	13.15
0.55	9.04	9.49	10.64	11.67	13.95	13.95	13.95	14.30	14.30	13.27
0.40	9.38	10.29	10.98	12.58	13.61	13.95	13.95	14.53	14.30	13.73
0.30	10.41	11.09	12.12	13.27	14.07	14.07	14.07	14.07	14.07	14.07
0.15	11.67	12.47	13.15	13.95	14.64	14.64	14.64	14.64	14.64	14.07
0.05	12.35	13.04	13.73	14.30	14.98	14.64	14.64	14.64	14.64	14.07

RPM

Figure 8.3 Lookup table for engine fuelling

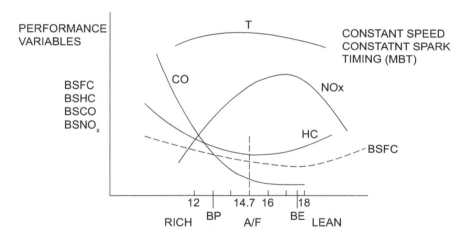

Figure 8.4 Effect of A/F (λ) on CO, BSFC, etc.

consumption, torque, HC, and CO. Slightly rich mixtures are required for cold engine starts, and weaker ones for the lowest brake specific fuel consumption. In practice the exhaust oxygen (λ) sensor plays a significant part in ECU control as it attempts to control fuelling at $\lambda = 1$. Atypical example of a 3 dimensional engine fuelling map is shown in Figure 8.5.

8.2 Factors affecting spark timing

Spark timing, required to achieve a set power, depends on three main factors such as speed, load and air-to-fuel ratio.

1 **Speed.** As the speed increases, the crank travels through a larger angle during the time taken for the gas to burn.

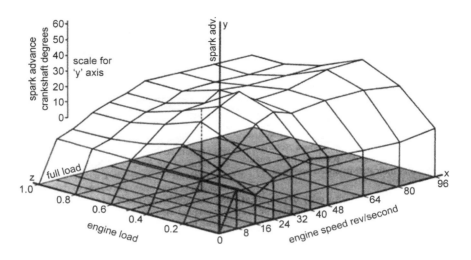

Figure 8.5 Ignition timing map

2   **Load.** As the load is increased, the opening of the throttle also increases to maintain a set speed. This increases the quantity of air and fuel entering the cylinder, which causes the higher compression pressure. As a result the flame rate increases and the gas burns more quickly.

3   **Air-to-fuel ratio.** A weaker mixture takes a longer time to burn than the chemically correct mixture.

Ideally the degree of ignition advance should be such that maximum cylinder pressure is achieved near to TDC, as shown in Figure 8.6. If the spark occurs too early, the maximum pressure is reached while the piston is rising on the compression stroke, and this can lead to knocking and mechanical damage. If the spark is retarded, the cylinder pressure is lower, and this means a lower power output. When the amount of spark advance is correct, the maximum pressure occurs shortly after TDC. For every operating condition of the engine there is an ideal angle for the spark to occur. The maximum brake torque occurs at an angle known as "the minimum advance for best timing" (MBT). Figure 8.7 shows that MBT produces best torque and minimum specific fuel consumption.

As with the fuelling map, the ignition test data is stored in ECU memory in the form of lookup tables that are compiled in computer language that is understood by the processor of the ECU microcontroller. The example of a

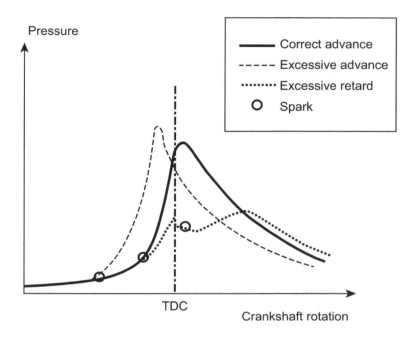

Figure 8.6 Effect of ignition advance on peak pressure

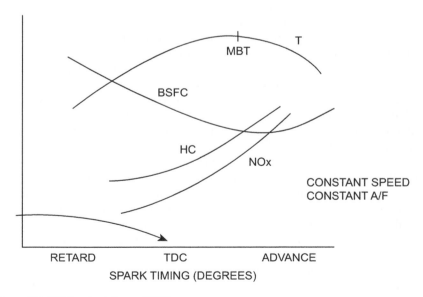

Figure 8.7 MBT and minimum BSFC

1995 5.0 Mustang Base Spark Table (Deg. BTDC)

	600	700	900	1100	1300	1500	2000	2500	3000	4000	5000
0.90	9.00	9.00	9.00	11.00	13.00	17.00	19.00	21.00	22.00	22.00	25.00
0.80	9.00	10.00	10.00	11.00	15.00	18.00	21.00	23.00	24.00	26.00	26.00
L 0.70	10.00	12.00	12.00	18.00	19.00	20.00	23.00	25.00	26.00	28.00	28.00
O 0.60	12.00	14.00	15.00	19.00	21.00	22.00	25.00	27.00	28.00	30.00	30.00
A 0.50	14.00	17.00	19.00	25.00	30.00	30.00	30.00	30.00	30.00	32.00	32.00
D 0.40	17.00	20.00	21.00	26.00	31.00	34.00	36.00	37.00	38.00	40.00	40.00
0.30	35.00	35.00	35.00	35.00	35.00	36.00	38.00	39.00	40.00	40.00	40.00
0.20	35.00	35.00	35.00	32.00	35.00	40.00	40.00	40.00	40.00	40.00	40.00
0.10	35.00	35.00	35.00	28.00	28.00	28.00	28.00	28.00	28.00	28.00	28.00

RPM

Figure 8.8 Ignition timing lookup table

Ford Mustang ignition map (Figure 8.8) shows the ignition advance angle for a range of engine speeds and loads. The left-hand scale of Load is probably a signal from the throttle position sensor and the bottom horizontal scale represents engine speed. For example, a throttle (Load) of 0.60 and engine speed of 1100 rev/min shows and ignition advance angle of 19°.

Diesel engine mapping

Buses used on urban routes where there is much stopping and starting require a different mapping from that of heavy haulage trucks where fuel economy, ability to maintain speed and high torque are essential features. A basic engine that is being mapped for buses will have a quite different map from the one for a heavy haulage vehicle. Vehicles that are equipped with power take-off (PTO) will need an engine mapping that takes account of the needs of other systems on the vehicle. Such a system is known as "intermediate speed control" that is initiated when PTO is in use. The process of individual engine mapping is known as vehicle related adaptation

The power output in a diesel engine is determined by the mass of fuel injected in each cycle – a small amount of fuel for idling and a larger amount for full power. Because the air drawn into the cylinders on each induction stroke is near constant, the air-to-fuel ratio varies greatly throughout the engine operating range, from around 60:1 at idling speed to around 12:1 at full power. The limits to air-to-fuel ratio are determined by the need to minimise smoke and other emissions and the need to maintain conditions for satisfactory combustion.

Figure 8.9 shows an engine map for a small diesel engine. This map shows how BMEP varies with brake specific fuel consumption which is indicated by the small letters on the curves.

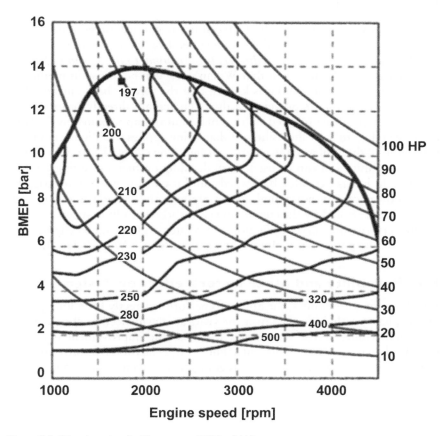

Figure 8.9 Diesel engine fuelling map – BSFC g/kWh

Example 8.1

A 4 cylinder, 4 stroke diesel engine developing a power output of 75 kW returns a brake specific fuel consumption figure of 200 g/kWh at a speed of 2400 rev/min. Calculate the amount of fuel injected at each injection operation.

Solution 8.1

Amount of fuel per hour $= 75 \times 200 = 15{,}000$ g

Number of injections $= \dfrac{\text{cylinders}}{2} \times \text{rpm} \times 60 = 288{,}000$ per hour

Amount of fuel per injection $= \dfrac{15{,}000}{288{,}000} = 0.052$ g $= 62$ mm^3

8.3 Willan's law

Willan's law was first used in the 19th century in connection with the power output of a throttle-controlled steam engine. *The law states that the indicated power of steam engines is directly proportional to the amount of steam used.* It has since been shown that Willan's law can be of use in predicting diesel engine behaviour, and it is known that the torque, brake mean effective pressure and brake power of a diesel engine are directly proportional to the amount of fuel injected up to a maximum of approximately 75% of full power. Figure 8.10 shows how Willan's line applies to a diesel engine. In the region of the Willan's line where a straight line exists, the data displayed is quite accurate and can be used when deciding on fuelling settings. When the straight line is extended backwards until it cuts the horizontal axis, the brake power on the horizontal axis represents the friction and pumping losses as represented by the engine power absorbed at idling speed.

The performance curves in Figure 8.11 relate to a large diesel engine, and the dot on the power curve shows the region in which Willan's law applies.

8.4 Reliability and other factors: effect on design

Reliability in automobiles, the ability to perform correctly for a long period of time, has greatly improved over the years as is evidenced by the existence of 100,000-mile, 7-year warranties. Much of the improvement is due to improved materials and lubricants, but a large factor of improvement is attributable to the designers. The foregoing details of engine mapping reveal

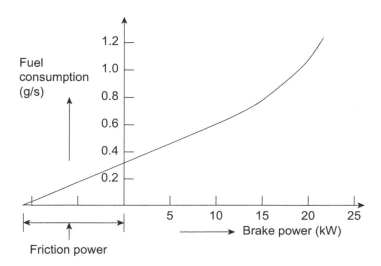

Figure 8.10 Willan's law

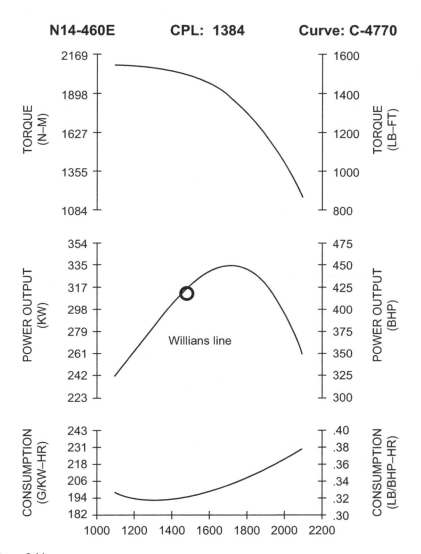

N14-460E **CPL: 1384** **Curve: C-4770**

Willians line

Figure 8.11

that it is a painstaking process that is undertaken by teams of engineers. Reliability of a system, over a long period of time, is an important consideration, and it may mean that a designer wishing to optimise power output may have to compromise over fuelling and other factors so as not to jeopardise the reliability. For example, it may be that fuelling could be increased at a certain engine speed to increase power output, at the expense of fuel economy and exhaust emissions. These are factors to be taken into account when considering altering the settings (remapping).

Self-assessment questions

8.1   Figure SAQ 8.1 shows the binary code for ignition advance angle. What angle of ignition advance is indicated in this lookup table?

Binary code ignition advance crank angle	Binary code Engine rpm
0 0 0 0 0 0 0 0	0 0 0 0 0 0 0 0
0 0 1 0 0 1 0 0	0 0 0 0 1 0 0 0
0 0 1 0 0 1 0 0	0 0 0 1 0 0 0 0
0 0 1 0 1 1 1 0	0 0 0 1 1 0 0 0
0 0 1 1 0 1 0 0	0 0 1 0 0 0 0 0
0 0 1 1 1 0 1 0	0 0 1 0 1 0 0 0
0 0 1 1 1 0 0 0	0 0 1 1 0 0 0 0
0 0 1 1 0 1 1 0	0 1 0 0 0 0 0 0
0 0 1 1 1 0 1 1	0 1 0 1 0 0 0 0
0 0 1 1 1 0 1 0	0 1 1 0 0 0 0 0

Ignition Output

Figure SAQ 8.1 Ignition lookup table

8.2   Figure SAQ 8.2 shows a lookup table for engine load, speed and air-to-fuel ratio. What is the air-to-fuel ratio at an engine load of 0.30, engine speed 1300 rev/min?

1995 5.0 Mustang Base OL Fuel Table (A/F Ratio)

LOAD	500	700	900	1100	1300	1500	2000	2500	3000	4000
0.90	8.46	8.01	9.49	10.29	12.35	12.35	12.35	12.01	12.01	12.01
0.80	8.46	8.46	9.95	10.52	12.92	13.15	13.50	13.61	13.61	12.92
0.70	8.81	8.81	10.41	10.98	13.73	13.95	13.95	14.30	14.30	13.15
0.55	9.04	9.49	10.64	11.67	13.95	13.95	13.95	14.30	14.30	13.27
0.40	9.38	10.29	10.98	12.58	13.61	13.95	13.95	14.53	14.30	13.73
0.30	10.41	11.09	12.12	13.27	14.07	14.07	14.07	14.07	14.07	14.07
0.15	11.67	12.47	13.15	13.95	14.64	14.64	14.64	14.64	14.64	14.07
0.05	12.35	13.04	13.73	14.30	14.98	14.64	14.64	14.64	14.64	14.07

RPM

Figure SAQ 8.2 Fuelling map

8.3 Which sensor in an engine management system generates the engine load signal?

8.4 What is the significance of the acronym MBT in relation to engine torque? Use fuel per stroke at given rpm, pick a value from the BSFC map and work out thermal efficiency.

8.5 Figure SAQ 8.5 shows that NOx are highest when the air-to-fuel ratio is slightly weak. How does EGR affect %NOx in the exhaust?

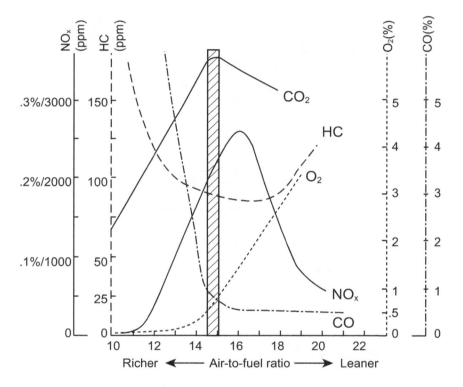

Figure SAQ 8.5 Emissions

8.6 The map in Figure SAQ 8.6 shows that the BMEP is 14 bar at 1800 rpm. It is a 4 cylinder, 4 stroke engine with a bore and stroke of 100 mm. Calculate:

(a) the brake power developed at this engine speed

(b) the brake thermal efficiency if the calorific value of the fuel is 44 MJ/kg; take BSFC as 200 gm/kWh.

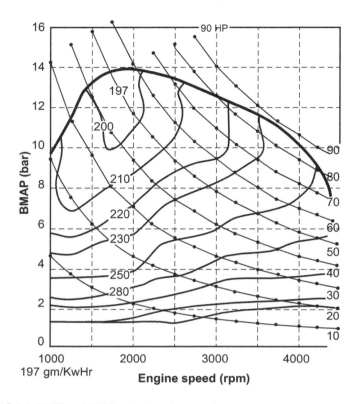

Figure SAQ 8.6 BMEP and BSFC against engine speed

Answer 8.6:

(a)   Brake power $= \dfrac{\text{BMEP} \times l \times a \times \text{rpm}}{60 \times 1000}$

$\text{bp} = \dfrac{14 \times 10^5 \times 0.1 \times 0.7854 \times 0.1 \times 0.1 \times 1800}{60,000} = 33\text{kW}$

(b)   Brake thermal efficiency $= \dfrac{\text{energy to power per second}}{\text{energy supplied in fuel per second}}$

$\text{BTE} = \dfrac{1000 \times 3600}{0.2 \times 44 \times 10^6} = 0.409 = 40.9\%$

Chapter 9

Fuels and other energy sources

9.1 Calorific value

The calorific value of a fuel is the amount of energy that is released by the combustion of 1 kg of the fuel. Two figures for calorific value of hydrocarbon fuels are normally quoted: these are the higher or gross calorific value, and the lower calorific value. These two values arise from the steam that arises from the combustion of hydrogen. The higher calorific value includes the heat of the steam, whereas the lower calorific value assumes that the heat of the steam is not available to do useful work. The calorific value that is quoted for motor fuels is that which is used in calculations associated with engine and vehicle performance. For example, the calorific value of petrol is approximately 44 MJ/kg. The approximate values of the properties of other fuels are shown in Table 9.1.

Table 9.1 Approximate values of fuel properties

Property	LPG	Petrol	Diesel oil	Methanol
Relative density	0.55	0.74	0.84	0.8
Useable calorific value	48 MJ/kg	44 MJ/kg	42.5 MJ/kg	20 MJ/kg
Octane rating	100	90	–	105
Chemical composition	82% C, 18% H_2	85% C, 15% H_2	87% C, 12.5% H_2, 0.5% S	38% C, 12% H_2, 50% O_2
Air-to-fuel ratio	15.5	14.7	14.9	6.5
Mass of CO_2 per kg of fuel used (kg)	3.04	3.2	3.2	1.4

9.2   The bomb calorimeter

The energy content of a fuel, which is known as the calorific value, can be determined from its chemical composition or by the use of a calorimeter, such as the bomb calorimeter which is described here.

The principle of the bomb calorimeter is that the heat released by the combustion of the fuel sample is the heat gained by the water and the calorimeter. As Figure 9.1 shows, there are two main parts of the apparatus. Part (a) shows the pressure vessel that is called the bomb. The ends of the bomb unscrew to permit the sample of fuel to be placed in the crucible. The amount of fuel is approximately 1 gram, the ignition fuse wire is also inserted at this stage and the bomb is reassembled. After reassembly the bomb is charged with oxygen at about 30 bar.

Part (b) shows the bomb in place inside the copper container. When this operation is completed, the copper vessel is filled with a measured quantity of water which is sufficient to completely cover the bomb. After ignition, the rise in temperature is carefully recorded until it ceases to rise and the final temperature is recorded. The water equivalent of the calorimeter is known as is the specific heat capacity of water. This description is intended as an overview; the actual procedure requires a great deal of care and attentions if

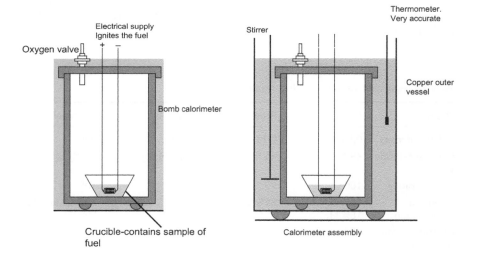

Figure 9.1 (a) and (b) The bomb calorimeter

a valid figure for calorific value is to be obtained. Example 9.1 shows some figures obtained from a properly conducted fuel test.

Example 9.1

Temperature rise $= 5.4$ C
Specific heat capacity of water $= 4.2$ kJ/kgC
Mass of fuel in the crucible $= 0.932$ g
Mass of water and water equivalent of calorimeter $= 1.93$ kG

Heat given up by fuel $=$ heat gained by water and calorimeter

mass of fuel $\times$ calorific value $=$ mass of water $\times$ specific heat of water $\times$ temp change

$$0.000932 \text{ kg} \times \text{calorific value} = 1.93 \text{ kg} \times \frac{4.2 \text{ kJ}}{\text{kg}} \times 5.4 \text{ C}$$

$$\therefore \text{calorific value} = \frac{1.93 \times 4200 \times 5.4}{0.000932} = 47 \text{ MJ} / \text{kg}$$

9.3   Combustion

Modern fuels for motor vehicles, such as petrol and diesel fuel, are principally hydrocarbons. That is to say, they consist largely of two elements, hydrogen and carbon. The proportions of these elements in the fuel vary, but a reasonably accurate average figure is that motor fuels such as petrol and diesel fuel are approximately 85% carbon and 15% hydrogen.

Products of combustion

When considering products of combustion, it is useful to take account of some simple chemistry relating to the combustion equations for carbon and hydrogen. An adequate supply of oxygen is required to ensure complete combustion, and this is obtained from the atmospheric air.

Relevant combustion equations

The relevant combustion equations are as follows:

For carbon, $C + O_2 = CO_2$. Because of the relative molecular masses of oxygen and carbon, this may be interpreted as: 1 kg of carbon requires 2.67 kg of oxygen and produces 3.67 kg of carbon dioxide when combustion is complete.

For hydrogen, $2H_2 + O_2 = 2H_2O$. Again, because of the relative molecular masses of H and O_2, this may be interpreted as: 1 kg of hydrogen requires 8 kg of oxygen and produces 9 kg of H_2O when combustion is complete.

9.4   Air-to-fuel ratio

Petrol has an approximate composition of 15% hydrogen and 85% carbon. The oxygen for combustion is contained in the air supply, and approximately 15 kg of air contains the amount of oxygen that will ensure complete combustion of 1 kg of petrol (see Table 9.2). This means that the air-to-fuel ratio for complete combustion of petrol is approximately 15:1; a more precise figure is 14.7:1, depending on the exact chemical composition of the fuel.

Total oxygen = 3.47 kg

Mass (weight) of air required per kg of fuel $= \dfrac{100}{23} \times 3.47 = 15.1 \ kg$

Air-to-fuel ratio for this fuel = 15.1:1.

9.5   Petrol engine combustion

Combustion in spark ignition engines such as the petrol engine is initiated by the spark at the sparking plug and the burning process is aided by factors such as combustion chamber design, temperature in the cylinder, mixture strength, etc. Because petrol is volatile, each element of the fuel is readily supplied with sufficient oxygen from the induced air to ensure complete combustion when the spark occurs. Petrol engine combustion chambers are designed so that the combustion that is initiated by the spark at the sparking plug is able to spread uniformly throughout the combustion chamber.

For normal operation of a petrol engine, a range of mixture strengths (air-to-fuel ratios) are required from slightly weak mixtures, say, from 20 parts air to 1 part of petrol for economy cruising, to 10 parts of air to 1 part of petrol for cold starting. During normal motoring, a variety of mixture strengths within this range will occur. For example, acceleration requires

Table 9.2 Oxygen and air per kg of fuel

Substance	Weight per kg of fuel	Oxygen needed per kg of constituent	Oxygen needed per kg of the fuel
Carbon (C)	0.85 kg	2.67 kg	0.85 × 2.67 = 2.27 kg
Hydrogen (H_2)	0.15 kg	8.00 kg	0.15 × 8 = 1.2 kg

approximately 12 parts of air to 1 part of petrol. These varying conditions plus other factors such as atmospheric conditions that affect engine performance lead to variations in combustion efficiency, and undesirable combustion products known as exhaust emissions are produced. Exhaust emissions and engine performance are affected by conditions in the combustion chamber, and two effects that are associated with combustion in petrol engines are detonation and pre-ignition.

Detonation

Detonation is characterised by a knocking and loss of engine performance. The knocking arises after the spark has occurred, and it is caused by regions of high pressure that arise when the flame spread throughout the charge in the cylinder is uneven. Uneven flame spread leads to pockets of high pressure and temperature that cause elements of the charge to burn more rapidly than the main body of the charge. Detonation is influenced by engine design factors such as turbulence, heat flow, combustion chamber shape, etc. Quality of the fuel, including octane rating, also has an effect. Detonation may lead to increased emissions of CO, NOx and HC.

Pre-ignition

Pre-ignition is characterised by a high pitched "pinking" sound that is emitted when combustion prior to the spark occurs, and it is caused by regions of high temperature. These regions of high temperature may be caused by sparking plug electrodes overheating, sharp or rough edges in the combustion region, carbon deposits and other factors. In addition to loss of power and mechanical damage that may be caused by the high pressures generated by pre-ignition, combustion may be affected, and this will cause harmful exhaust emissions.

9.6 Octane rating

The octane rating of a fuel is a measure of the fuel's resistance to knock. A high octane number indicates a high knock resistance. Octane ratings are determined by standard tests in a single cylinder, variable compression ratio engine. The research octane number (RON) of a fuel is determined by running the test engine at a steady 600 rpm while the compression ratio is increased until knock occurs. The motor octane number (MON) is determined by a similar test, but the engine is operated at higher speed. The research octane number (RON) is usually higher than the MON, and fuel suppliers often quote the RON on their fuel pumps. An alternative rating that is sometimes used gives a figure which is the average of RON and MON.

9.7   Diesel fuel

Diesel fuel has a calorific value of approximately 45 MJ/kg and a specific gravity of approximately 0.8 g/cc. The ignition quality of diesel fuel is denoted by the Cetane number, and a figure of 50 indicates good ignition properties. Among other properties of diesel fuel that affect normal operation are flash point, pour point and cloud point, or cold filter plugging point.

Flash point

The flash point of a fuel is the lowest temperature at which sufficient vapour is given off to cause temporary burning when a flame is introduced near the surface. A figure of 125°F (52°C) minimum is quoted in some specifications. The flash point is determined by means of apparatus such as the Pensky Martin flash point apparatus, and the test is performed under controlled laboratory conditions.

Pour point

The pour point of a fuel is the temperature at which a fuel begins to thicken and congeal and can no longer be poured. A pour point of zero degrees F or −18°C is suitable for some conditions.

Cloud point

The cloud point, sometimes known as the cold filter plugging point (CFPP), is the temperature at which the fuel begins to have a cloudy appearance and will no longer flow freely through a filtering medium. Cloud point occurs at a temperature of approximately 20°F (11°C) above the cloud point temperature.

Note

These figures for diesel fuel are approximate and presented here as a guide only. Readers who require more detailed information are advised to contact their fuel supplier.

9.8   Exhaust emissions

The exhaust gases are the products of combustion. Under ideal circumstances, the exhaust products would be carbon dioxide, steam (water) and nitrogen. However, owing to the large range of operating conditions that engines experience, exhaust gas contains several other gases and materials such as those in the following list;

- CO – due to rich mixture and incomplete combustion
- NOx – due to very high temperature
- HC – due to poor combustion
- PM – soot and organo-metallic materials
- SO_2 – arising from combustion of small amount of sulphur in diesel fuel.

Factors affecting exhaust emissions

During normal operation, the engine of a road vehicle is required to operate in a number of quite different modes as follows:

Modes of road vehicle engine operation:

- Idling – slow running
- Coasting
- Deceleration – overrun braking
- Acceleration
- Maximum power.

These various modes of operation give rise to variations in pressure, temperature and mixture strength in the engine cylinder, with the result that exhaust pollutants are produced.

Hydrocarbons (HC)

The HC appear in the exhaust as a gas, and it arises from incomplete combustion due to a lack of oxygen. The answer to this might seem to be to increase the amount of oxygen to weaken the mixture. However, weakening the mixture gives rise to slow burning, combustion will be incomplete as the exhaust valve opens and unburnt HC will appear in the exhaust gases.

Idling

When the engine is idling, the quantity of fuel is small. Some dilution of the charge occurs because, owing to valve overlap and low engine speed, scavenging is poor. The temperature in the cylinders tends to be lower during idling and this leads to poor vaporisation of the fuel and HC in the exhaust.

Coasting, overrun braking

Under these conditions, the throttle valve is normally closed. The result is that no or very little air is drawn into the cylinders. Fuel may be drawn in from the idling system. The closed throttle leads to low compression pressure and very little air. The shortage of oxygen arising from these conditions causes incomplete combustion of any fuel that enters the cylinders, and this results in HC gas in the exhaust.

Acceleration

Examination of torque versus specific fuel consumption for spark ignition engine reveals that maximum torque occurs when the specific fuel consumption is high. Because the best acceleration is likely to occur at the maximum engine torque, a richer mixture is required in order to produce satisfactory acceleration. Fuelling systems provide this temporary enrichment of appropriate increases in mixture strength to meet demands placed on the engine. This leads to a temporary increase in emissions of HC and CO. As the engine speed rises, combustion speed and temperature increase, and this gives rise to increased amounts of NOx.

High speed, heavy load running

Here the engine will be operating at or near maximum power. Examination of the power versus specific fuel consumption shows that maximum power is produced at higher specific fuel consumption figures and richer mixtures. Increased emissions of CO and HC are likely to result.

Cruising speed – light engine load

Under these conditions, where the engine is probably operating in the low specific fuel consumption speed, the mixture strength is likely to be around 16:1 or higher to provide good fuel economy. Other emissions are lower under these conditions.

9.9   European emissions standards (Euro 6)

Euro 6 emissions standards (petrol)
CO: 1.0 g/km
THC: 0.10 g/km
NMHC: 0.068 g/km
NOx: 0.06 g/km
PM: 0.005 g/km (direct injection only)

Euro 6 emissions standards (diesel)
CO: 0.50 g/km
HC + NOx: 0.17 g/km
NOx: 0.08 g/km
PM: 0.005 g/km
PN [#/km] 6.0×10^{11}/km PN [#/km]: 6.0×10^{11}/km (direct injection only)

Oxides of nitrogen (NOx)

Oxides of nitrogen (NOx) are formed when combustion temperatures rise above 1800°K.

Hydrocarbons (HC)

Unburnt hydrocarbons arise from:

- unburnt fuel remaining near the cylinder walls after incomplete combustion being removed during the exhaust stroke
- incomplete combustion due to incorrect mixture strength.

Carbon monoxide (CO)

Carbon monoxide is caused by incomplete combustion arising from lack of oxygen.

Sulphur dioxide (SO$_2$)

Some diesel fuels contain small amounts of sulphur which combines with oxygen during combustion. This leads to the production of sulphur dioxide, which can, under certain conditions, combine with steam to produce H_2SO_3, which is a corrosive substance.

Particulate matter (PM)

The bulk of particulate matter is soot which arises from incomplete combustion of carbon. Other particulates arise from lubricating oil on cylinder walls and metallic substances from engine wear.

Carbon dioxide (CO$_2$)

Whilst CO_2 is not treated as a harmful emission, it is thought to be a major contributor to the green-house effect, and efforts are constantly being made to reduce the amount of CO_2 that is produced. In the United Kingdom, the amount of CO_2 that a vehicle produces in a standard test appears in the vehicle specification so that it is possible to make comparisons between vehicles on this score. The figure is presented in grams per kilometre. For example, a small economy vehicle may have a CO_2 figure of 145 g/km and a large saloon car a CO_2 figure of 240 g/km. Differential car tax rates are applied to provide incentives to users of vehicles that produce smaller amounts of CO_2.

9.10   Alternative fuels

It is now generally accepted that the world's oil resources are finite and that they are being depleted at a rapid rate. Attention is concentrating on alternative fuels and methods of propulsion for motor vehicles. Alcohols such as methyl alcohol, or methanol as it is commonly known, are produced from vegetable matter. Bio-fuels are said to be environmentally valuable because their products of combustion are, very roughly, water (H_2O) and CO_2. It is argued that the CO_2 from the combustion of these fuels is consumed by the vegetation that is producing the crop that will make the next supply of fuel – this process is referred to as a "closed carbon cycle".

Methanol is not strictly a hydrocarbon because it contains some oxygen. The calorific value of methanol is approximately 26 MJ/kg. Methanol has a higher latent heat value than petrol, and it has higher resistance to detonation. Whilst the higher latent heat value and relatively high ignition temperature of methanol indicate that higher compression ratios can be used, a disadvantage is that vaporisation at low temperatures is poor and this can lead to poor cold starting ability.

Liquified petroleum gas (LPG)

Petroleum gases such as butane and propane are produced when oil is refined to produce liquid fuels such as DERV and petrol. Once liquefied and stored under pressure, LPG will remain liquid until is exposed to atmospheric pressure. The chemical composition of propane is C_3H_8 and butane is C_4H_{10}, the relative densities at 15°C are approximately 0.5 and 0.57 respectively, while the calorific value is slightly higher than petrol at approximately 46 MJ/kg. A considerable industry exists to support the conversion of road fuelling from petrol and DERV to LPG.

In some countries including the United Kingdom, favourable tax systems support the use of LPG.

Hydrogen

Compressed hydrogen may be stored on a vehicle and used in an internal combustion engine. Among the advantages claimed for it are no carbon dioxide emissions and products of combustion that are primarily water. The main future use of hydrogen as a propellant for vehicles is thought to be as a source of energy in fuel cells. The electricity produced in the fuel cell is used as a power source for the electric motor that replaces the internal combustion engine of the vehicle. A simple fuel cell is shown in Figure 9.2. The fuel cell consists of two electrodes, an anode and a cathode, that are separated by an electrolyte. Hydrogen passes over hydrogen acts on the anode and oxygen from the atmosphere acts on the cathode. The

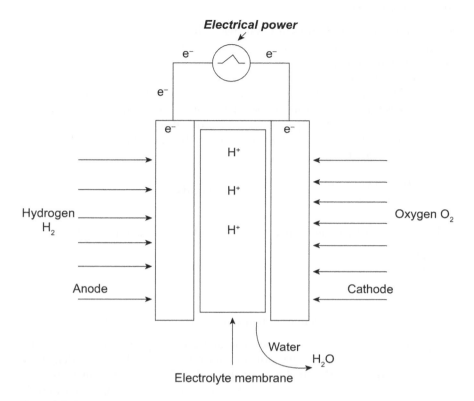

Figure 9.2 The hydrogen fuel cell

catalytic action of the anode causes the hydrogen atom to form a proton and an electron. The proton passes through the polymer electrolyte to the cathode, and the electron passes through an external circuit to the cathode. This action provides an electric current in external circuit. In the process the hydrogen and oxygen combine to make water which is the principal emission.

9.11   Zero emissions vehicles (ZEVs)

Operation of fuel cells for vehicle propulsion does not involve combustion in an engine and the normal products of combustion and associated pollutants are not produced. The main product of the electro-chemical processes in the fuel cell is water; consequently, vehicles propelled by fuel cells and electric motors are known as zero emissions vehicles.

Supercapacitors

Supercapacitors are often used in electric vehicles for temporary storage of electrical energy during regenerative braking operations because they have very high capacitance.

As shown in Figure 9.3, a capacitor is two conductors (usually metal) separated by a material known as a dielectric, and its ability to store an electric charge is known as its **capacitance**. The unit of capacitance is the farad (F), and it is defined as the capacitance of a capacitor that requires a p.d. of 1 volt to maintain a charge of 1 coulomb on that capacitor. This definition allows us to state that:

$$\frac{\text{charge in coulombs}}{\text{p.d. in volts}} = \text{capacitance in farads}$$

If Q = number of coulombs, V = p.d., and C = capacitance, then;

$$\frac{Q}{V} = C$$

$$Q = CV.$$

Capacitance can be calculated from the formula $\varepsilon_0 \varepsilon_r \dfrac{A}{d}$, where $\varepsilon_0 \varepsilon_r$ are the permittivities of free space and the dielectric respectively. A is the surface area and d is the thickness of the dielectric material. It is evident that the surface area and the thickness of the dielectric are important factors in the amount of capacitance, and this is a factor in considering their suitability for use in motor vehicle systems.

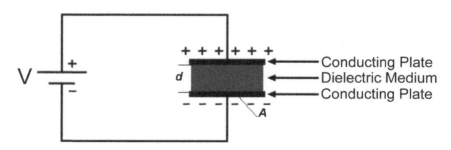

d = distance between plates metres

A = total area of plates in sq metres

Figure 9.3 A simple capacitor

Example 9.2

A capacitor with a capacitance of 90 microfarads is connected across a DC supply of 450 V. Calculate the amount of electric charge that is stored in the capacitor.

Solution

$$\text{Quantity of charge } Q = \frac{90}{10^6} \times 450 = 0.045 \text{ coulomb}$$

Note that this is a very small amount of electricity, because a coulomb is a flow of 1 ampere for a period of 1 second, and for vehicle use much larger quantities of electrical energy are required which has led to the development of super (ultra) capacitors which can store large amounts of electrical energy. The unit of electrical energy that is normally used in connection with electric vehicles is the kilowatt hour (kWh) which is equivalent to 3.6 MJ.

In order to increase the area of capacitance material, the capacitor may be of the type of construction where a large surface area is provided in a small space.

9.12   Capacitors in parallel

Example 9.3

In Figure 9.4 both C_1 and C_2 are subjected to the same p.d. of V volts. The charge on the capacitors is:

$$Q_1 = C_1 V \text{ and } Q_2 = C_2 V$$

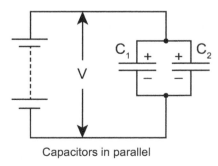

Capacitors in parallel

Figure 9.4

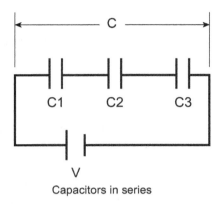

Capacitors in series

Figure 9.5 Capacitors in series

If C_1 and C_2 are replaced by a single capacitor of C farads so that the total charge of $(Q_1 + Q_2)$ is produced by the same p.d., then $Q_1 + Q_2 = CV$. If we now substitute for Q_1 and Q_2, we get the following: $C_1V + C_2V = CV$.

Divide through by V to produce $C = C_1 + C_2$ farad.

9.13 Capacitors in series

When capacitors are connected in series, the total capacitance is found as follows

$$\frac{1}{C} = \frac{1}{C_1} + \frac{1}{C_2} + \frac{1}{C_3} + - - -$$

Example 9.4

Three capacitors with capacitances of 2, 4 and 8 microfarads respectively are connected in series. Determine the total capacitance C.

$$\frac{1}{C} = \frac{1}{2} + \frac{1}{4} + \frac{1}{8} = 0.875\mu F$$

$$\therefore C = \frac{1}{0.875} = 1.143\mu F$$

Connecting capacitors in series reduces the effective capacitance but increases the voltage of a pack.

9.14   Charging and discharging capacitors

Figure 9.6 shows the way in which the p.d. of the capacitor increases as it is charged from a DC source. A point is marked on the graph where the capacitor is charged to 63.2% of the maximum value. The time taken to reach this point (state of charge) is known as the time constant of the capacitor, and it is given by the formula $t = RC$, where R is the resistance in ohms and C is the capacitance in farads. It is known that it takes five time constants for the capacitor to be fully charged, and this is shown on the graph. For discharge, it takes one time constant for the charge to drop to 36.8 of the fully charged p.d. Both of these factors are important with regard to super-capacitors for use in vehicle systems because they affect the way in which they can be utilised.

9.15   Energy stored in a capacitor

Figure 9.7 is a graph that shows the behaviour of a capacitance of C farads being charged at a constant rate of I amperes for a period of t seconds and the charge placed in the capacitor is It coulombs, where I is the current in

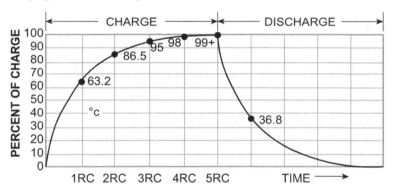

Figure 9.6 Capacitor time constant

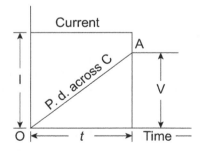

Figure 9.7 Charging a capacitor at constant rate

amperes and t is the time in seconds. The final p.d. across the capacitor is V volts, which means that the charge is also CV coulombs.

$$\therefore CV = It \, and \, V = \frac{It}{C}$$

This shows that the p.d. across C is proportional to the time t and is represented by the straight line OA.

The average p.d. across C during charging $= \frac{1}{2}V \, volts$, and the average

power to C during charging $= I \times \frac{1}{V} \, watts$

$\therefore$ average energy to C during charging $= avergage \, power \times time$

$$= \frac{1}{2}IVt \, joules$$

$$= \frac{1}{2}V \times CV \, joules$$

The electrostatic energy stored in $C = \frac{1}{2}CV^2 \, joules$.

Example 9.5

A capacitor with a capacitance of 3 farads is fully charged to its terminal voltage of 398 volts. Calculate the amount of energy stored in it. If this energy is fed into a vehicle propulsion system in a time of 6 seconds, how much extra power does it represent?

Solution 9.5

Energy stored in capacitor $= \frac{1}{2} \times 398^2 \times 3 = 0.238 \, Mj$.

$$\text{Power} = \frac{\text{energy}}{\text{time}} = \frac{238000}{6 \times 1000} = 39.7 \, kW$$

9.16 Supercapacitors

Supercapacitors are called "super" because they store much larger quantities of energy than conventional capacitors, and they achieve this feature by virtue of the materials used and the type of construction, which are shown in Figure 9.8.

Electric Double Layer Capacitors (EDLCs) do not have any dielectrics in general, but rather utilize the phenomena typically referred to as the

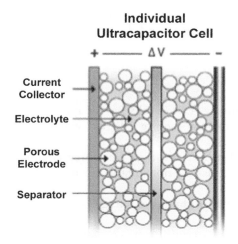

**Individual
Ultracapacitor Cell**

+ ——— Δ V ——— –

Current
Collector

Electrolyte

Porous
Electrode

Separator

Internal Cell Construction

Figure 9.8 Tecate group supercapacitor

electric double layer. In the double layer, the effective thickness of the "dielectric" is exceedingly thin, and because of the porous nature of the carbon the surface area is extremely high, which translates to a very high capacitance. However, the double layer capacitor can only withstand low voltages (**typically less than 3V per cell**), which means that electric doublelayer capacitors rated for higher voltages must be combined in **matched series-connected** individual capacitors, much like series-connected cells in higher-voltage batteries.

(Tecate group)

Advantages

- **High energy storage.** Compared to conventional capacitor technologies, EDLCs possess orders of magnitude higher energy density. Very suitable for regenerative braking, stop start and other short burst of energy applications.
- **Low Equivalent Series Resistance (ESR).** Compared to batteries, EDLCs have a low internal resistance, hence providing high power density capability.
- **Low temperature performance.** Compared to batteries, EDLC's are capable of delivering energy down to –40°C with minimal effect on efficiency.

- **Fast charge/discharge.** Since EDLCs achieve charging and discharging through the absorption and release of ions and coupled with its low ESR, high current charging and discharging is achievable without any damage to the parts.

Disadvantages

- **Low per cell voltage.** EDLC cells have a typical voltage of 3 V. Since many applications may require a higher voltage, the cells have to be connected in series.
- **Cannot be used in AC and high frequency circuits.** Because of their time constant, EDLCs are not suitable for use in AC or high-frequency circuits.
- **Can provide only short bursts of energy.**

Ultracapacitors have a typical time constant of approximately one second. One time constant reflects the time necessary to charge a capacitor 63.2% of full charge or discharge to 36.8% of full charge. This relationship is illustrated in Figure 9.6 A typical time constant of 1 second is quoted for supercapacitors which means that they are charged and discharged in about 5 seconds.

Example 9.6

A regenerative braking system operates at 312 volts DC and it is supported by a capacitor bank of 144 capacitors connected in series. If each capacitor has a capacitance of 350 farads, calculate the capacitance of the bank.

Solution 9.6

If the total capacitance is

$$C_{tot}, \quad \frac{1}{C_{tot}} = \frac{1}{C_1} + \frac{1}{C_2} + \cdots \frac{1}{C_n}, \text{ where } n = \text{number of capacitors}$$

then

$$C_{tot} = \frac{350}{n} = \frac{350}{144} = 2.431 \text{ farads}$$

9.17   Batteries

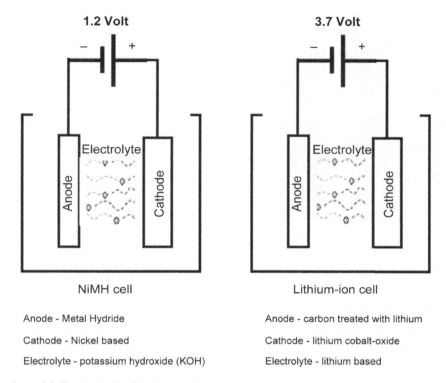

1.2 Volt   **3.7 Volt**

NiMH cell   Lithium-ion cell

Anode - Metal Hydride   Anode - carbon treated with lithium

Cathode - Nickel based   Cathode - lithium cobalt-oxide

Electrolyte - potassium hydroxide (KOH)   Electrolyte - lithium based

Figure 9.9 Some details of battery packs

NiMH battery

A typical NiMH electric vehicle battery contains a number of packs of cells called modules. Each module contains eight cells to give a module voltage of 9.6 volts. A number of these modules are connected together in a battery pack to provide the total voltage for a particular vehicle. In some vehicles this is 288 volts, while in others it is 144 volts.

Li-ion battery

In some applications of li-ion batteries for electric vehicles, 10 of the 3.6 volt cells are grouped into a "stack" to give a stack voltage of 36 V. These stacks are then joined together in a battery pack to give the required voltage of 144 V, or 288 V.

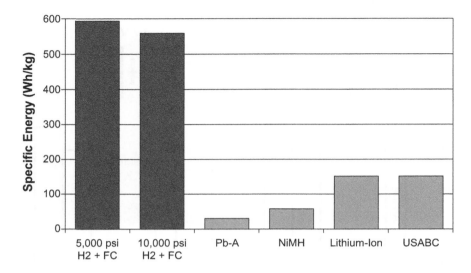

Figure 9.10 Relative specific energy of fuel cells and batteries

Specific energy

The specific energy of a cell is the amount of electrical energy in Watt hours (Wh) per kg of battery weight. It is a measure of the amount of electrical energy that can be stored in each cell, and it is a figure that is used when making comparisons between battery types. Figure 9.10 shows how the energy storage capacity of various batteries and fuel cells varies and highlights how the hydrogen fuel cell capacity is much higher than all of the batteries.

State of charge (SOC)

The state of charge of the battery pack is constantly monitored by an ECU so that regenerative braking can be controlled to suit battery SOC and the driver to be made aware of the amount of motoring that is available at any time.

9.18   Moment of inertia of a uniform disc

When a disc such as a flywheel is rotating, the whole of its mass may be considered to be placed in a ring at a radius that is the radius of gyration k (see Figure 9.11). The moment of inertia of a flywheel may be calculated from the formula $I = mk^2$, where I is the moment of inertia, m is the mass of the flywheel in kg and k is the radius of gyration in metres. For a simple disc flywheel, the radius of gyration is $k = \dfrac{r}{\sqrt{2}}$ where r is the outer radius of the flywheel.

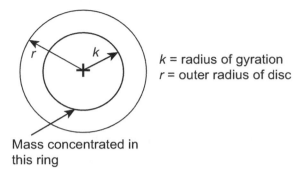

Mass concentrated in
this ring

Figure 9.11 Moment of inertia *I* of a disc

Example 9.7

A simple disc flywheel has a radius of 0.30 m and a mass of 50 kg. Determine its moment of inertia.

Solution 9.7

$$I = mk^2 = 50 \times \left(\frac{0.30}{1.414}\right)^2 = 50 \times 0.045 = 2.25\,\text{kgm}^2$$

Kinetic energy stored in a rotating flywheel

Kinetic energy is the energy that an object possesses as a result of its velocity. In the case of an object moving in a straight line with a velocity of V m/s, the kinetic energy is $KE = \frac{1}{2}mV^2$, where m is the mass in kg and V is the velocity in m/s. In the case of the flywheel where the mass is concentrated in a ring at a radius of k, the linear velocity of the mass is $V = \omega k$, where ω is the angular velocity of the flywheel in radians per second. If we now replace V by ωk in the kinetic energy formula as follows:

$$KE = \frac{1}{2}mk^2\omega^2 = \frac{1}{2}I\omega^2 \text{ joules}$$

Example 9.8

A flywheel of mass = 8 kg and radius of gyration 150 mm is rotating at 3600 rev/min. Determine its kinetic energy.

The kinetic energy of the rotating flywheel $= \frac{1}{2}I\omega^2$.

$I = mk^2 = 8 \times 0.15 \times 0.15 = 0.18 \text{ kgm}^2$

Angular velocity $\omega = \dfrac{3600}{60} \times 2\pi = 377 \text{ rad/s}$

Flywheel kinetic energy $= \dfrac{1}{2}(0.18) \times 377 \times 377 = 12792$ joules

KERS (kinetic energy regeneration system)

The use of a flywheel to store energy from the slowing down of a vehicle and then use the energy to help accelerate the vehicle has been in use on trains for many years. In recent times a great deal of attention has been drawn to this use because of the KERS system that is used in Formula 1 racing. In most vehicle braking systems, the kinetic energy of the vehicle is converted into heat energy at the brakes by virtue of the friction. In regenerative systems, the linear kinetic energy of the vehicle is converted into rotary kinetic energy in the flywheel system, or into electrical energy in battery or super capacitor systems.

Broadly speaking, regenerative braking systems are of two types:

Flywheel systems, where the linear kinetic energy of the vehicle is converted into rotary kinetic energy of a flywheel,
Electrical systems, where a generator-motor is used to charge batteries or super capacitors.

In both cases the intention is to use the inertia of the system to assist in slowing the vehicle down, and the braking effect of the flywheel stores the kinetic energy that derives from the mass and speed of the vehicle. When required, the flywheel energy is transmitted to the drive system of the vehicle to provide extra power for acceleration.

The flywheel system

This is called the kinetic energy restoration system (KERS). Figure 9.12 shows the outline details of a system that is used by Volvo cars. The system consists of a rotating flywheel, a continuously variable transmission system (CVT), a step up gearing between the flywheel and the CVT and a clutch which connects this system to the final drive of the transmission. When the brakes are applied, or the vehicle decelerates, the clutch connecting the flywheel system to the driveline/transmission is engaged, causing energy to be transferred to the flywheel via the CVT. The flywheel stores this energy as rotational energy and can rotate up to a maximum speed of 60,000 rpm.

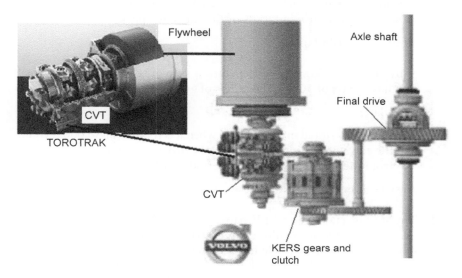

Figure 9.12 Volvo KERS (Courtesy Volvo Cars)

When the vehicle stops, or the flywheel reaches its maximum speed, the clutch disengages the flywheel unit from the transmission, allowing the flywheel to rotate independently. Whenever this stored energy is required, the clutch is engaged and the flywheel transmits this energy back to the wheels, via the CVT. Generally the flywheel can deliver up to 60 kW of power.

Example 9.9

On cars the KERS system may be designed to hold 400 kJ that derives from the deceleration of the vehicle. If 300 kJ is put back into the drive train to accelerate the vehicle in a time of 5 seconds, an additional power of power = (change of energy)/(time taken) = (300 kJ)/(5 s) = 60 kW is provided.

9.1 A petrol fuel has a composition of 14% hydrogen and 86% carbon by mass.

Determine:

(a) the ideal (stoichiometric) air-to-fuel ratio for this fuel
(b) the mass of CO_2 produced per Kg of fuel used.

Answer 9.1: (a) 14.87:1, (b) 3.16 kg

9.2 A certain battery-electric-car has a supercapacitor pack in addition to the batteries. The electrical power system operates at 280V DC and the supercapacitor pack can store 2 kWh when fully charged. Determine the capacitance in farads of the capacitor pack.

Answer 9.2: 184 farads

9.3   A flywheel kinetic energy regeneration system (KERS) has a flywheel that has a mass of 10 kg. The radius of gyration of the flywheel is 6 cm.

Calculate:

(a)   the energy stored in the flywheel when it is rotating at 50,000 rev/min
(b)   the extra power introduced into the drive system if it is transferred from the KERS flywheel in a period of 10 seconds.

Answer 9.3: 49.37 kW

9.4   The specific energy of a BEV battery is given in kWh/kg and the total energy available in a fully charged battery is given in kWh. A certain battery electric vehicle has a battery capacity of 40 kWh and the specific energy is 0.22 kWh/kg. In average motoring, the vehicle consumes electrical energy at a rate of 0.250 kWh/km.

Calculate:

(a)   the weight of the BEV battery
(b)   the theoretical range of the vehicle.

Answer 9.4: (a) 181.8 kg, (b) 160 km

9.5   An engine on test consumes 2.2 kg of mixture at a temperature of 25°C and the temperature of the exhaust gas was 800°C. Calculate the amount of energy carried away in the exhaust gas every minute. Take the specific heat capacity of the exhaust gas as 1.05 kJ/kgC.

Answer 9.5: 1790 kJ/min

9.6   On a dynamometer test, a four cylinder four stroke engine returns a brake specific fuel consumption of 0.3 kg/kWh. The calorific value of the fuel is 43 MJ/kg.

Calculate:

(a)   the brake thermal efficiency
(b)   the mass of fuel injected on each power stroke.

Answer 9.6: (a) 26%, (b) 0.833 mg per injection

9.7   Describe the difference between higher and lower calorific values of a hydrocarbon fuel. The following results were recorded when a bomb calorimeter was being used to determine the calorific value of a liquid fuel:

- Mass of fuel burned = 0.515 g
- Mass of water in calorimeter = 2420 g
- Water equivalent of calorimeter = 400 g
- Initial temperature of water and calorimeter = 14°C
- Final temperature of water and calorimeter = 16°C.

Calculate the calorific value of the fuel.

Answer 9.7: 46 Mj/kg

9.8   Compare exhaust gas recirculation and selective catalyst reduction as two alternative methods of reducing NOx emissions. Pay particular attention to initial costs, weight of equipment, maintenance factors, effect on engine power, availability of urea as a reagent, and any other factors that you think should be considered.

9.9   Conduct research to satisfy yourself about the effect on fuelling requirements that arise when considering the merits and demerits of converting a spark ignition engine to run on alcohol fuel. Pay particular attention to the calorific value and general effect on engine performance.

9.10 What are the advantages of hydrogen as a motor fuel in respect of atmospheric pollution?

Chapter 10

Electric propulsion

An overview

10.1 Why electric?

Throughout the 20th century the internal combustion engine and the automobile evolved together. Initially there were few vehicles, and it was not until after the First World War that vehicle ownership began to become widespread. In the second half of the century, as the number of vehicles in use grew, concern about their effect on the environment began to emerge. One of the first concerns was about smoke emission from diesel engines and hidden emissions such as lead from tetra-ethyl lead in petrol. As awareness grew of global warming and the way CO_2 emissions contributed to it, measures were introduced to reduce the amount coming from motor vehicles. More recently, the harmful effect of oxides of nitrogen (NOx) and other emissions such as particulate matter (PM) have become better understood. Measures to eliminate such emissions have not been entirely successful, and several countries are now considering stopping the manufacture of internal combustion engine vehicles. Most major motor vehicle manufacturing countries now have programmes in place for large-scale production of electrically powered vehicles. This chapter covers some of the technology that constitutes the powertrain of an all-electric vehicle, as opposed to a hybrid where an additional internal combustion engine is also used. All-electric vehicles are sometimes referred to as BEVs (battery electric vehicle).

Figure 10.1a shows how vehicle usage in the USA increased throughout the twentieth century, this pattern of vehicle usage has happened on a global scale.

Figure 10.1b shows how carbon dioxide in the atmosphere has increased. Most of this increase is attributed to vehicle emissions

10.2 Characteristics of electric motors that make them suitable for vehicles

Figure 10.2 clearly shows some significant differences between the performance of IC engines and electric motors. For example, the electric motor produces high torque at low speed, whereas the IC engine produces its

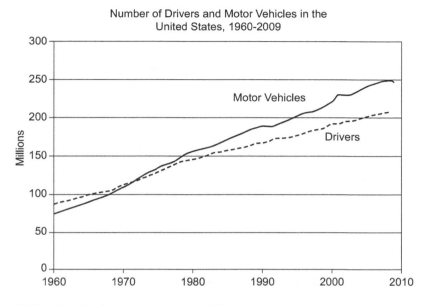

Figure 10.1(a) Growth of vehicle population in USA

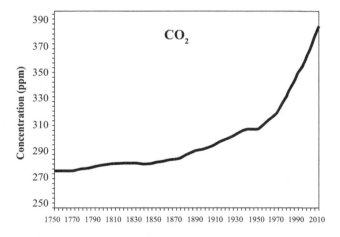

Figure 10.1(b) Growth of CO_2 emissions worldwide

highest torque around the mid-speed range. This means that with electric propulsion, it is not necessary to slip the clutch in order to produce a smooth pull away from a stationary position. Closely related to this is improved gradeability at low speed. After the constant torque range, the

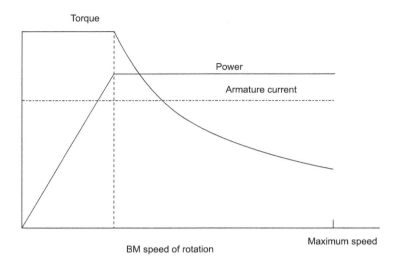

Electric motor power and torque

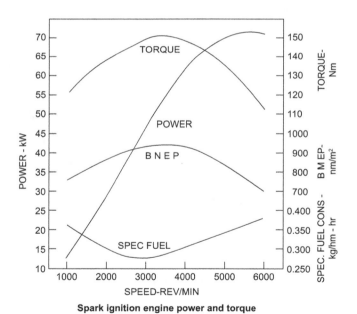

Spark ignition engine power and torque

Figure 10.2 Electric propulsion versus the internal combustion engine

electric motor control switches to the constant power mode which continues throughout the speed range of the motor. This part closely relates to the ideal tractive effort curve for a vehicle in motion. Compare this to the IC engine, where the power changes until a maximum value occurs near the maximum speed of the engine. The near perfect shape of the electric motor power curve suggests that gear changes may not be necessary. In terms of efficiency of energy conversion it is difficult to make a direct comparison, because in an IC engine vehicle, the conversion from chemical energy in the fuel into mechanical energy to propel the vehicle takes place in the vehicle itself; whereas in an electrically propelled vehicle, the energy conversion takes place in the power station. The thermal efficiency of a petrol engine is approximately 25%, while that of a diesel engine is about 35%. In an electric vehicle there are few losses in the conversion from electrical energy into mechanical energy in the vehicle itself, and it is reasonable to suggest that the thermal efficiency of the electric vehicle relates to the thermal efficiency of the power station, which is said to be approximately 40%. An electric vehicle causes very little atmospheric pollution in the places where it is used, whereas pollution from the conversion from chemical to mechanical energy in an IC engine vehicle occurs in the places where that vehicle is used. The arguments about this point of view may be many and varied, but they are not pursued here. Various methods of comparing the performance of electric vehicles with IC engines are used; one of these is called the "Well to Wheel" comparison. In this method, an assessment is made of the cost of all the processes that are involved in getting the crude oil from the well, refining it and distributing it. Other methods use such measures as a drive cycle test at the manufacturing stage. It is probably the case that most owners are interested in cost per mile, reliability, range and performance, and recharging facilities.

Position of electric motors on vehicle

(a) Front wheel drive – needs a differential.
(b) Conventional 4 wheel drive – needs a differential front and rear.
(c) Front wheel drive, separate motor each side, no differential. A disadvantage of this type of propulsion motor is the greatly increased unsprung mass.
(d) All wheel drive, no differential.

Where the wheels are motorised on each side of the axle, the differential gear may be dispensed with because the computerised control system will ensure that wheels on either side can rotate at different speeds, as is necessary when the vehicle is cornering. Electronic regenerative braking is available only on those wheels that are driven by an electric motor.

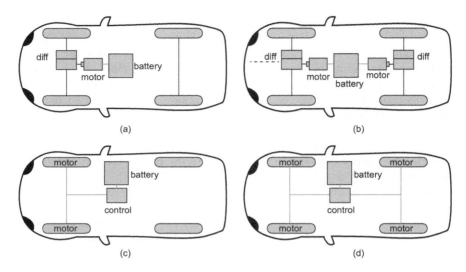

Figure 10.3 Positions of motor generator on the vehicle

10.3   Direct current motors

In Figure 10.4 the magnetic field is produced by permanent magnets which produce flux density of B Webers/m^2. When the armature coil is rotated, it cuts through the lines of magnetic force, and this produces an electromotive force (EMF voltage) in the armature; the machine is therefore acting as a generator. When voltage is applied at the commutator, current flows in the armature, the magnetic effect that surrounds each coil of the armature reacts with the permanent magnetic field and produces rotation of the armature, and this is the electric motor. In Figure 10.4 a force F is shown as acting on the armature, and this force acting at the radius of the armature multiplied by the radius of the armature coil produces a torque T.

The force $F = $ BIL Newtons

The force F acts on both sides of the coils and the torque produced is

Torque $T = F \times d$ Newton metres

In an actual motor there are many coils (windings), and this equation is useful for explaining how torque and speed control of the motor can be achieved. For example, the magnetic effect B is often produced by an electromagnet, the strength of which is controlled by the electric current flowing

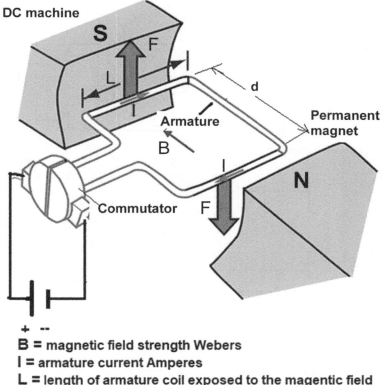

DC machine

B = magnetic field strength Webers
I = armature current Amperes
L = length of armature coil exposed to the magentic field
F = force exerted on armature Newtons

Figure 10.4 General principle of electric motor/generator

through the magnetic field winding. The I is the electric current supplied to the armature, and that can be varied. The L in the equation relates to the length of the armature windings that are subject to the magnetic flux B. L cannot be altered because it is feature of the motor's construction.

Types of DC motor generator

Any of the DC machines shown in Figure 10.5 can be used as a motor or a generator. The one that is most suitable for BEV traction is the separately excited machine, because the armature current and the field current can be controlled separately. The variable resistor control of current is not satisfactory for BEV use, and electronic current control is used instead.

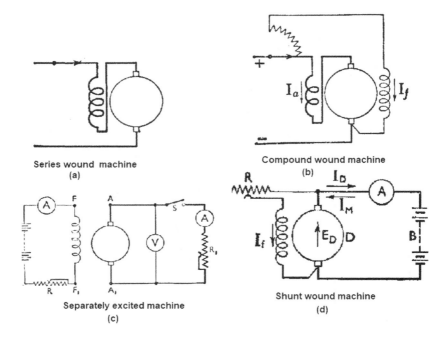

Figure 10.5 Types of DC motor generator

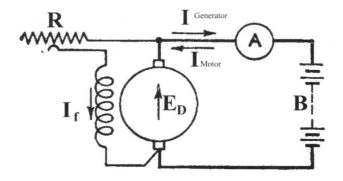

Figure 10.6 Circuit of a simple DC machine

Motor-generator

I_d = dynamo or generator current
I_m = motor current
I_f = field current always in the same direction

The relationship between the current, the e.m.f., the voltage V and resistance of the armature can be shown as follows:

E = e.m.f. is generated in the armature
V = terminal voltage
R_a = resistance of the armature circuit
I_a = current through the armature

When the machine is operating as a generator:

$$E = V + I_a R_a$$

When the machine is operating as a motor, the e.m.f., E, is less than the applied voltage V, and the direction of the armature current is the reverse of that when the machine is a generator. This gives us:

$$E = V - I_a R_a$$
$$V = E + I_a R_a$$

Because the e.m.f. generated in the armature of a motor is in opposition to the applied voltage, it is often called *back e.m.f.*

Example 10.1

The armature of a DC machine has resistance of 0.1 ohm and it is connected to a 200 V supply. Calculate the generated e.m.f. when it is running (a) as a generator producing an armature current of 80 A, (b) when it is running as a motor with an armature current of 60 A.

Solution 10.1

(a) 208 V, (b) 194 V.
    Solution (a) generator

$$E = V + I_a R_a$$
$$V = 200, \ I_a = 80, \ R_a = 0.1$$
$$E = 200 + 80 \times 0.1 = 208 \, \text{Volts}$$

(b) Motor $E = V - I_a R_a$
    $$V = 200 - 6 = 194 \, \text{Volts}$$

10.4   The brushless direct current motor (BLDC)

This type of machine operates on alternating current principles and it is preferred by some BEV manufactures because if offers advantages over other types, such as:

- good speed versus torque characteristics
- high efficiency
- good dynamic response
- long operating life (no brushes)
- relatively quiet in operation
- good speed ranges.

Figure 10.7 shows the basic construction of a typical electric motor generator.

Principle of BLDC

The coils in the stator are arranged in sets called phases. Each phase is energised by an electric charge from the chopper circuit. The magnetic field that is created reacts with the magnets in the rotor, causing it to rotate. In order to do this, the rotor magnets and the stator magnetic field must be correctly aligned, and this is the function of the Hall type sensors. When the Hall sensor senses that the rotor and stator field are correctly positioned, it sends a signal to the chopper, which then directs electric current to the appropriate phase windings.

The Hall sensors

BLDCs usually have three sensors (see Figure 10.8) that give a high or low signal indicating that a N or S magnetic pole is passing the Hall element.

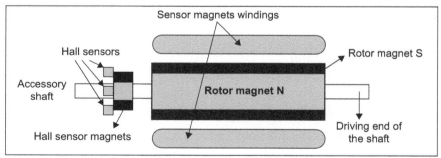

Section along the length of BLDC machine

Figure 10.7 BLDC machine (source microchip technology)

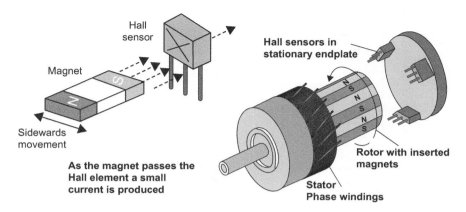

Figure 10.8 The Hall sensors

10.5 Motor control

Performance characteristics of BLDC

The Hall sensors send a signal to the controller which uses a pulse width signal to control the size of the current in each phase and this determines the speed and torque of the motor. The direction of phase current is shown in Figure 10.9, both in the respective table and by the arrows in the diagram.

Reverse drive

Table (b) in Figure 10.9 shows that a different sequence for energising the phases produces reverse. (Note the reverse sequence is not shown in the phase diagram.)

Pulse width

The power output of the motor is controlled by input from the accelerator sensor which is processed by the microcontroller which determines the size of the duty cycle (Figure 10.10).

The BLDC motor generator (Figure 10.5) operates in four modes that are shown in the four quadrants of Figure 10.11.

Traction motor control

Control of the traction motor is performed by chopper circuits, and a slightly different chopper circuit is used for motor control in each of the

Phase sequence forward drive

(a)

Phase current		
A	B	C
DC+	Off	DC-
DC+	DC-	Off
Off	DC-	DC+
DC-	Off	DC+
DC-	DC+	Off
Off	DC+	DC-

Phase sequence reverse drive

(b)

Phase current		
A	B	C
Off	DC-	DC+
DC+	DC-	Off
DC+	Off	DC-
Off	DC+	DC-
DC-	DC+	Off
DC-	Off	DC+

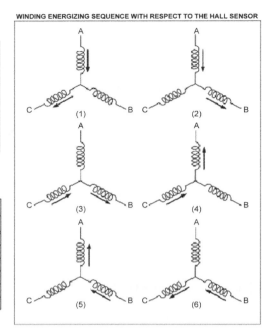

Figure 10.9 The sequence of phase current for motor control

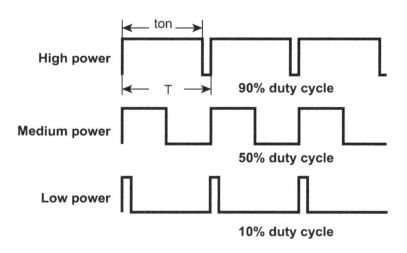

Figure 10.10 Duty cycle for power output

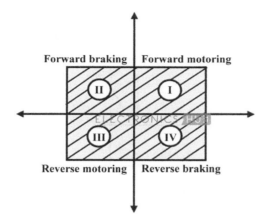

Figure 10.11 The four modes of operation

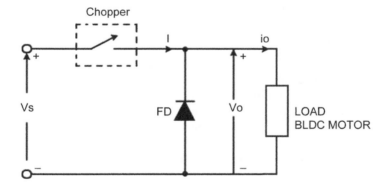

Figure 10.12 Forward motion chopper circuit

quadrants. A chopper is a switch that operates between on and off at a frequency determined by the microcontroller which is a major component of the control system.

Motor control in the four quadrants

First quadrant

In this quadrant (Figure 10.12), the chopper switch is placed in series with the motor and the diode FD protects the circuit as the chopper switch is opened. During this stage of operation the chopper receives direct current

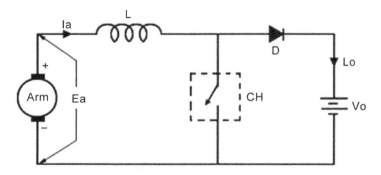

Figure 10.13 Forward regeneration braking

from the battery and converts it to chopped dc which acts like alternating current in the motor. The duty cycle determines the power output in accordance with the microcontroller program and other sensors including accelerator (load speed) demand.

Second quadrant

Figure 10.13 shows a chopper in regenerative braking mode. When the chopper is switched on, the motor-generator electrical energy is stored in the inductor L, and when the chopper is switched off, the energy is released. When Ea exceeds Vo, the motor acts as a generator and current flows in the opposite direction to when it was operating as a motor, this recharges the battery pack. The effort (torque) required to operate the generator is transferred to the driving wheels to provide the braking effect.

Third and fourth quadrants

In these quadrants the control unit changes the motor control system so that it rotates in the reverse direction. The chopper circuits are similar to those used to control the motor in quadrants 1 and 2.

Regen braking and battery charging

In vehicles propelled by an IC engine, a considerable amount of braking when slowing down for a corner or descending a hill is obtained by engaging a lower gear and closing the throttle. The engine is then driven by the vehicle, and the resulting braking effect is felt at the driving wheels and increases any torque that is applied by the vehicle's friction brakes. There is little of this effect with electric motor propulsion apart from a small amount to overcome inertia of the rotating parts. In order to overcome this factor and to make use of the kinetic energy of the moving vehicle that is to be converted during the braking process, the electric propulsion motor is switched to generating mode. The vehicle then drives the electric motor, and the torque and power required to produce electricity is provided by the vehicle, the momentum of which is driving the vehicle, and this manifests itself as a considerable braking torque at the driving wheels. The braking energy now produced by the motor-generator is used to recharge the battery pack, as opposed to an IC engine vehicle braking where the braking kinetic energy is dissipated as heat. Because the kinetic energy of the moving vehicle is now converted into electricity which recharges the battery the term regenerative braking is used. In addition to the battery pack, the system may also be equipped with super capacitors to temporarily store regen electrical energy.

Regeneration

In vehicles propelled by an IC engine, a considerable amount of braking when slowing down for a corner or descending a hill is obtained by engaging a lower gear and closing the throttle. The engine is then driven by the vehicle and the resulting braking effect is felt at the driving wheels and increases any torque that is applied by the vehicle's friction brakes. There is little of this effect with electric motor propulsion apart from a small amount to overcome inertia of the rotating parts. In order to overcome this factor and to make use of the kinetic energy of the moving vehicle that is to be converted during the braking process, the electric propulsion motor is switched to generating mode. The vehicle then drives the electric motor and the torque and power required to produce electricity is provided by the vehicle, the momentum of which is driving the vehicle, and this manifests itself as a considerable braking torque at the driving wheels. The braking energy now produced by the motor-generator is used to recharge the battery pack, as opposed to an IC engine vehicle braking where the braking kinetic energy is dissipated as heat. Because the kinetic energy of the moving vehicle is now converted into electricity which recharges the battery the term regenerative braking is used. In addition to the battery pack, the system may also be equipped with super capacitors to store electrical energy above the amount that the battery can accept.

Control of regeneration

The extent of regen braking must be controlled to prevent skidding and to take account of varying road conditions. Example 10.2 uses figures taken at random, but they demonstrate the braking effect. The downward load is affected by the deceleration rate and the friction force by tyre and road condition. If the regen braking effect is too high, as detected by the ABS sensor, the regen controller must reduce it. Another factor in the control of regen is the state of charge (SOC) of the battery. If the SOC is of the order of 90%,

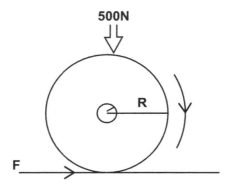

Figure 10.14 Braking effect

regen is cut off by reverting to the coasting mode. Figure 10.14 shows how frictional grip affects braking ability.

Example 10.2

F = friction force opposing wheel spin

μ = coefficient of friction between tyre and road = 0.7

R = radius of wheel = 0.3 m

Force acting down = 500 N

Braking torque = $F \times R = 0.7 \times 500 \times 0.3 = 105$ Nm

To control the level of braking the PWM duty cycle is varied, which essentially toggles the current flow between regeneration and coasting. The maximum level of regeneration occurs when the current flow is as shown in Figure 10.15. At the lower end when no regeneration is required, the current flow is as shown in Figure 10.16. To control the level of regen braking the microcontroller toggles between coasting and full regeneration.

Electric vehicle performance

The data shown in Figure 10.17 relates to a 55 kw electric motor, and it is typical of the performance of the electric motors that are used in motor vehicles. For the purposes of the following example, the data in the figure is used to show vehicle performance, namely:

1   Acceleration
2   Maximum speed
3   Gradeability.

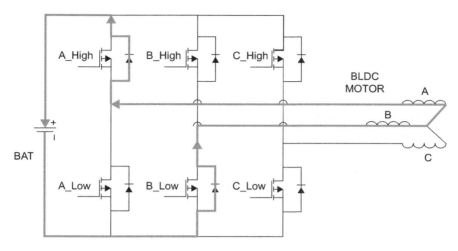

Figure 10.15 Regeneration current flow

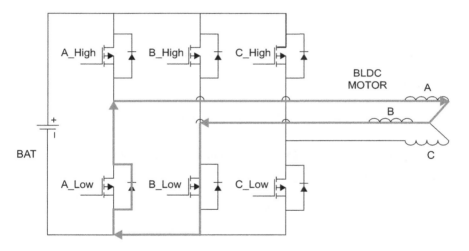

Figure 10.16 No regeneration

Example 10.3

An electrically propelled car weighs 1250 kg, the rolling radius of the wheel and tyre is 0.28 m and the final drive has a reduction ratio of 7.9:1. Using data from Figure 10.17, determine:

1 Maximum acceleration.
2 Maximum possible speed in km/h.

3   The gradeability at traction motor speeds of 1000 rev/min and 9000 rev/min. At the higher speed, the aerodynamic and rolling resistance amount to a force of 500 N.

Take $g = 9.81$ m/s^2 and transmission efficiency $= 95\%$.

Solution 10.3

1   *The maximum acceleration will occur in the region where the torque is highest.*

The maximum torque is 120 Nm from start-up to 4000 rev/min motor speed, and this is where maximum acceleration will occur.
The torque after the final drive reduction of 7.9:1 $= 120 \times 7.91 \times 0.95 = 901$ Nm
The tractive effort (force) at the driving wheels $\dfrac{\text{wheel torque}}{\text{radius of wheel}} = \dfrac{901}{0.28}$
$= 3218$ N.

Tractive force $=$ vehicle mass $\times$ acceleration

$$\therefore acceleration = \frac{tractive\,force}{vehicle\,mass} = \frac{3218}{1250} = 2.57 m/s^2.$$

2   *Speed of the vehicle = revs per hour of wheel × circumference of wheel.*

$$\text{revs of wheel at} 1000 \text{ rpm of motor} = \frac{12000}{7.9} \times 60 = 91140 \text{ rev / hour}$$

$$\therefore \text{speed of vehicle} = 91140 \times 0.56 = 160,363 \frac{\text{m}}{\text{h}} = \frac{160.4\,\text{km}}{\text{h}}.$$

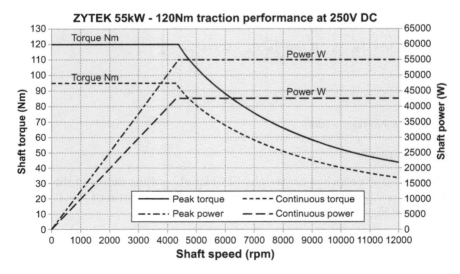

Figure 10.17 Performance -torque and power.

3   *From the graph, the torque at 1000 rpm the motor torque = 120 Nm*

The torque at the road wheel $= 120 \times 7.9 \times 0.95 = 901$ Nm

The tractive effort $= \dfrac{\text{wheel torque}}{\text{radius of wheel}} = \dfrac{901}{0.28} = 3218$ N

Gradeability

Force pushing up slope = slope force + neglect wind and rolling resistance

Tractive effort $= Mg \sin\theta$

$\sin\theta = \dfrac{Te - 500}{Mg} = \dfrac{3218}{1250 \times 9.81} = \dfrac{3218}{12,263} = 0.262$

Gradient $\theta = 15.2$ degrees

From the graph at 9000 rpm of the motor, torque = 45 Nm

The tractive effort at 9000 rpm $= \dfrac{45 \times 7.9 \times 0.95}{0.28} = 1206$ N

Self-assessment questions

10.1 A battery electric vehicle weighing 1400 kg has its speed reduced from 60 km/h to 40 km/h under a regenerative braking operation. Calculate the amount of energy that is available to recharge the battery.

Answer 10.1: 109 kJ, 0.03 kWh

10.2 The maximum tractive T_e available at the road wheels is given by the formula;

$$T_e = \dfrac{T_{m.}OGR\eta_t}{r}$$

where
T_m = maximum torque of the drive motor
OGR = the overall gear ratio
η_t = efficiency of motor and drive train
r = rolling radius of the road wheels.

Calculate the maximum tractive effort at the road wheels of an all-electric car where the maximum drive motor torque is 200 Nm, the OGR is 5:1, the rolling radius of the road wheels is 0.30 m and the overall efficiency of the drive train is 85%.

Answer 10.2: 2.833 kN

10.3 Determine the maximum speed of an electric car where the maximum speed of the electric drive motor is 5000 rev/min, the overall gear ratio is 3:1 and the rolling radius of the road wheel and tyre is 0.28 m.

Answer 10.3: 176 km/h.

10.4 Determine the gradeability of an electric car weighing (mass) 1300 kg if the overall gear ratio is 6:1, the rolling radius of the tyre and wheel is 0.30 m and the maximum drive motor torque is 140 Nm with a powertrain efficiency of 90%. Take wind and rolling resistance and the effect of angular inertia as a force of 500 N.

Answer 10.4: 9 degrees

10.5 A battery electric car (BEV) weighing 1.4 tonnes has a drive motor torque of 200 Nm at a road speed of 72 km/h. The powertrain efficiency is 80%, the overall gear ratio is 5:1 and the rolling radius of the tyre and wheel is 0.31 m. The car has a drag coefficient $C_d = 0.30$ and an effective frontal area of 2.0 m². Determine the maximum acceleration at this speed, on a level road. Take rolling resistance as 250 N.

Answer 10.5: max acceleration = 0.55 m/s²

10.6 Compare the energy consumption of a petrol driven car that returns an average fuel consumption 6 litres/100 km with a similarly sized electric car that has an energy consumption of 0.3 kWh/km. What will be the estimated range of the electric car if the battery capacity is 40 kWh?

Answer 10.6: Petrol car 0.536 kWh/km, electric car 0.30 kWh/km. Range of electric car = 133 km.

10.7 Using this table, compare the total energy content of NiMh and Li-Ion batteries each weighing 700 kg.

Answer 10.7: NiMh capacity 56 kWh. Li-Ion 105 kWh.

Index

Note: Numbers in **bold** indicate a table. Numbers in *italics* indicate a figure.

For Product Safety Concerns and Information please contact our EU representative GPSR@taylorandfrancis.com Taylor & Francis Verlag GmbH, Kaufingerstraße 24, 80331 München, Germany

Printed and bound by CPI Group (UK) Ltd, Croydon, CR0 4YY
01/05/2025
01858483-0001